WITHDRAWN FROM
KENT STATE UNIVERSITY LIBRARIES

Southern Illinois University Press
Series in
Aviation Management

David A. NewMyer, Editor

Aviation Industry Regulation
Harry P. Wolfe and David A. NewMyer
1985   ISBN: 0-8093-1177-1

Aviation Maintenance Management

Frank H. King

Southern Illinois University Press
Carbondale and Edwardsville

Library of Congress Cataloging in Publication Data

King, Frank H., 1918–
　Aviation maintenance management.

　(Southern Illinois University Press series in
aviation management)
　Includes bibliographical references and index.
　1. Airplanes—Maintenance and repair—Management.
I. Title. II. Series.
TL671.9.K56　1985　　　　629.134'6'068　　　　85-1983
ISBN 0-8093-1210-7

Copyright © 1986 by the Board of Trustees, Southern Illinois University
All rights reserved
Printed in the United States of America
Designed by Dan Gunter
Production supervised by Kathleen Giencke

89　88　87　86　　4　3　2　1

Contents

Figures vii
Tables ix
Preface xi

1. Introduction 1
2. The Federal Aviation Administration 12
3. Regulatory Requirements 30
4. Organizational Structures 50
5. Management Responsibilities 66
6. Aviation Maintenance Procedures 78
7. Applications of Aviation Maintenance Concepts 93
8. Budgeting, Cost Controls, and Cost Reduction 106
9. Training and Professional Development in Aviation Maintenance 114
10. Safety and Maintenance 122
11. Electronic Data Processing 145
12. Aviation Maintenance Management Problem Areas 152
13. Forecast and Summary 160

Appendixes

A. FAA Advisory Circulars—Maintenance (Abbreviated Listing) 167
B. References to FAA Reporting Requirements and Forms 177
C. Glossary 191
   Notes 197
   Index 205

Figures

1.	Type certification process	15
2.	1983 Piper Dakota	16
3.	FAA regional boundaries	23
4.	FAA rulemaking process	31
5.	Inspection signoff–return to service	40
6.	Boeing 727-100	43
7.	Boeing 747 and Boeing 707	44
8.	Maintaining continuing aircraft airworthiness	45
9.	Production airplane certification process	46
10.	Initial maintenance program development	47
11.	Ongoing maintenance program	49
12.	Example of functional organization	51
13.	Example of line organization	51
14.	The Boeing Company organization chart	52
15.	Maintenance and engineering organization of a typical major airline	53
16.	Generalized airline maintenance organization	54
17.	United Airlines maintenance operations	56
18.	United Airlines maintenance facility	57
19.	Inertial navigation equipment test	58
20.	Inside the exhaust funnel	59
21.	Maintenance hangar for United Airlines' wide-bodied aircraft	60
22.	High-intensity lights	61
23.	General Electric CF-6 engine inspection	62
24.	Inert gas welding operation	63

25.	Organizational structure of a typical commuter airline	64
26.	Typical fixed base operator organizational chart	65
27.	First page of scheduled maintenance form	80
28.	FAA operations specifications form	101
29.	Boeing 737-200 "C" check	102
30.	Engine maintenance	103
31.	Break-even graph	107
32.	Assembled General Electric CJ-805 turbofan engine	120
33.	Fire triangle	131
34.	Restricted areas during engine operations	132
35.	Typical aircraft ground service connections	142
36.	Aircraft maintenance history	147
37.	Aircraft status report	147

Tables

1. Scope of Airline Industry Employment in the United States (Section 401 Carriers) — 2
2. United States Air Carriers' Scheduled Service and Accident Statistics, 1973–1983 — 7
3. Development of United States Air Transportation — 8
4. Passenger Enplanements at the 30 Busiest United States Airports in 1983 — 9
5. Aircraft in Service (United States Scheduled Airlines) — 10
6. United States Airline and General Aviation Fleets — 11
7. Types of Inspections — 100
8. Passenger Fatalities per 100 Million Passenger Miles (Five-Year Averages) — 123
9. Growth of United States Aviation — 161

Preface

AVIATION MAINTENANCE MANAGEMENT is basically intended for college students majoring in aviation curriculums. The contents pertain to all facets of maintenance management concerns that confront the various levels of aviation supervision and the text fills a vital void in the field of aviation maintenance management.

Each chapter is devoted to a better understanding of the relationships between the Federal Aviation Administration and its regulations and reguirements and those segments of the aviation industry that must adhere to the FAA policies in the field of maintenance. All in all, the book represents a practical and understanding approach to aviation maintenance management problems.

The initial chapter highlights an introduction to aviation by discussing the impact of population changes on the airline industry and the importance of airline safety. Subsequent chapters cover the FAA organization structure, FAA publications pertaining to maintenance, the application of aviation maintenance management concepts, and other management responsibilities that deal directly with the maintenance supervisory function. It goes without saying that a book of this nature requires data from many segments of the federal government and the aviation industry. Without these vital and important inputs, the author could not have accomplished what he set out to do—write an overall comprehensive maintenance management text for aviation students.

The author would like to express sincere appreciation to the following for the excellent material and suggestions received that went into the publication of this text: American Airlines, Aviation Information Services, Beech Aircraft Corporation, Bell Helicopter Textron, Cessna Aircraft Company, Douglas Aircraft Company—Facilities Planning and Product Support, Frontier Airlines, Gates Learjet Corporation, General Aviation District Office II, National Transportation Safety Board, Piper Aircraft Corporation, Stits Poly-Fiber Aircraft Coatings, The Boeing Company, and United Airlines. And a special thanks to the following individuals for their contributions to this book: Carolyn Acord, Charles "Smokey" Doyle, David NewMyer, and Art Thompson.

Aviation Maintenance Management

1
Introduction

AVIATION MAINTENANCE ACTIVITIES are the backbone of all successful aviation enterprises. Good maintenance provides safer and more reliable aircraft, increases aircraft usage, and provides confidence of air travel to the approximately 300 million passengers that want to enjoy the freedom, timeliness, and safety of modern aircraft. Good maintenance management is that tangible asset that provides for the aviation industry the essentials necessary to the establishment of flying confidence by the public. Without good maintenance management, the aviation enterprise is adversely affected.

The maintenance manager is continuously encountering a variety of problems involving phases of technology, administration, production, personnel, and management functions. In order to direct the maintenance effort of his or her organization, the manager must understand the required operating principles. The complexity of general aviation and commercial airline aircraft has dramatically increased through the years, and as a result, the new airframe and powerplant (A&P) technician is finding it more and more difficult to learn the skills of maintaining and inspecting aircraft. The guiding of this effort thus requires more planning and expertise on the part of the maintenance manager. His or her program must be dynamic rather than static, which is to say it must operate to accommodate changes in industry needs, student needs, and instructional processes.

For years it has been recognized that the aviation industry is a vital adjunct to the growth of our nation. Today's aviation input consists of over 12,700 airports, 2,830 scheduled airline aircraft, and 215,000 general aviation aircraft.[1] The industry services both the industrial and the public sectors, ranging in scope from passenger-carrying aircraft, to cargo freight, business, air taxi, rental and commuter operations, special purpose, personnel, sport, and instructional flying.

In terms of personnel, the scheduled airlines averaged 328,648 (Table 1) employees in 1983.[2] General aviation in 1982 shows a figure of 250,000 broken down as follows: 70,000 people in sales and service support for commuter operations, flight training, maintenance, and systems support; 25,000 engaged in agricultural flying; 35,000 in corporate flying and support; 10,000 in industrial special uses; 10,000 self-employed instructors and mechanics; and 100,000 engaged in the manufacture of aircraft components (airframe, engines, instruments) and subcomponents.[3]

Few citizens of this country realize that in June 1978 a new population shift of historic proportions had come into being. Statistically then, for the first time in the nation's history, more people lived south of the Mason-Dixon line than north of it. As of the moment, even if you were to combine all the industrial centers of the North—Boston, Chicago, Cleveland, New York, Milwaukee, Pittsburgh—all of them and all the people between—you

TABLE 1
SCOPE OF AIRLINE INDUSTRY EMPLOYMENT IN THE UNITED STATES
(SECTION 401 CARRIERS)

	1973	1982	1983
Pilots and copilots	27,192	28,144	26,997
Other flight personnel	7,567	6,900	6,581
Flight attendants	42,819	50,860	53,535
Communications personnel	1,948	838	821
Mechanics	47,049	43,393	38,798
Aircraft and traffic servicing personnel	90,193	87,813	91,319
Office employees	59,891	66,997	67,382
All others	34,840	35,564	30,217
Total majors and nationals		320,509	315,650
Total large and medium regionals		9,986	12,998
Total employment	311,499	330,495	328,648

Source: *Air Transport, 1984: The Annual Report of the U.S. Scheduled Airline Industry* (Washington, D.C.: Air Transport Association of America, June 1984), p. 7.

would find that the population majority had still shifted to the South and Southwest.

The combination of data terminals located thousands of miles from the "Home Office," and a national air transportation system, offered a level of communication to provide the impetus for a massive and continuing relocation at a rapid pace of the nation's industrial base. This relocation was an integral part of the population shift. As a result, the United States has created a national air transportation system that is the envy of the world. While providing hundreds of thousands of jobs in both production and operations, it is highly contributory to dominance in the world's air routes—one of the few industries in such a position that this nation still enjoys.

The combination of public demand for lower fares and federal government deregulation has initiated a program that is already resulting in drastic change to the popular concept viewed as "the air lines" or the major scheduled "trunk carriers."[4] To get low fares, the traveler will get less flights. Literally to survive, by 1985 and beyond, the trunk carriers will continue to drop off more and more of the marginal or unprofitable stops with less such options available to travelers. Many of the traditional routes and schedules of the airlines are being eliminated or reduced. The situation unfolding mandates a major expansion of the regional and commuter air carriers (with the commuter airlines operating general aviation aircraft) both to fill the void and expand the service demanded by a mobile and growing population. Simply put, the major segments of air transportation—trunk carriers and general aviation—are coming closer together in purpose and consistently greater cooperation as the only way they can move their common and respective customer, the traveling public.

The major segments of the national air transportation system have no choice: they will become more interdependent. As the trunk and regional airlines reduce flights, general aviation (consisting of commuter carriers and private business aircraft) must expand. The moderate or long distance traveling public will have little choice; the greater efficiency of passenger mile per pound of fuel consumed will dictate that this public travel by air, or it may possibly not travel at all.

While power plants, trains, buses, homes, factories, even automobiles may possibly convert to use various alternate forms of energy, air transportation for the next quarter century will survive only with liquid petroleum. The 1980s will be the years that fact becomes more prominent. For the major carriers, fuel has gone from 12 percent of overall expenses in 1971, to over 30 percent in 1981. In just one year, from 1978 to 1979, the fuel bill for United States' scheduled airlines has gone from $4.1 billion to an estimated $6.5 billion.

Higher fuel costs mean higher seat break-even factors, and that means less flights to many communities. It becomes apparent, then, that the combination of commuter and private business flying that makes up general aviation is becoming an ever-increasing participant in the national air transportation system. A study by the National Business Aircraft Association shows one-third of all flights to airports with scheduled air carriers are to make passenger connections. Airline pasengers have no choice in the matter; the trunk carriers and general aviation will become more integrated. But the major difficulty facing air travelers will be lack of enough gates to load or unload at major hubs where the segments come together.

It is wasteful to require a 747, which may consume 10 to 20 thousand pounds of fuel per hour, to be held in the airport landing pattern. Similarly, a small airplane which may use as little as 60 to 100 pounds per hour doesn't need the long runways. Kennedy International Airport is a perfect example—with long parallel runways for constant flow of two large aircraft simultaneously, plus a small third runway for small aircraft—of what can be done in this area.

It is currently accepted that maintenance managers play a key role in today's aviation industry, whether they are in the airline sector, general aviation, or the federal government. In today's aviation environment, the manager is confronted with recognizing the need for academic, cultural, and vocational emphasis in a rapidly changing world, and a changing society. The manager must be, of necessity, dynamic, an organizer, and a high-caliber technical-scientific individual. Management understands that any devised program that he or she must initiate will be both technical and general in nature, must be dynamic, must be organized rather than fragmented, and must be technological so as to take advantage of such modern technology as the latest in high-speed computers, robotics as used in production processes, and other scientific advancements.

The educational systems, being what they are today, are facing some severe concerns in the areas of subject matter, curriculum emphasis, and curriculum completion date requirements. For example, the FAA has established curricula and time completion dates of 1,900 hours (A&P certificate) to be completed within a 24-month period. However, a recent study proposes some minor changes to FAA Regulation 147 (Aviation Maintenance Technical Schools) by recommending greater technical depth and subjects which meet the general educational requirements for university credit. Even though it is conservatively estimated that there are 1.5 million to 2.0 million pople working in the civilian aviation industry, little is known about the overall employment possibilities in this industry. This is particularly true at the high school and community college vocational education levels where more could be done to train people for the aviation industry.[5]

There has been some concern in recent years regarding the number of clock hours required by subject area to meet FAA certification requirements. It is being recommended that, with continued improvements through curriculum development programs and the introduction of new educational technologies in future training programs, certification emphasis will be on student achievement rather than on course hours of instruction.

A professional set of performance criteria tests will have to be developed which would be used to determine whether the student should be certificated for all or any part of the overall curriculum. If properly developed, the performance criteria tests could be used for "subcertification," that is, as a welder, sheet metal specialist, hydraulic specialist, electronic specialist, and so on. Each of these specialty certificates would be added to and become a part of a person's overall A&P license.

The above stated concept would permit a variety of flexible actions on the part of the student, the training institutions, and the certification/accreditation bodies, as follows:

1. The student who is not able to attend a two or four year program, but must, for economic reasons, seek employment at the earliest possible time, can obtain his or her training in a specialty field, such as welding, and be certificated by the FAA. The person can then become productive in the aviation maintenance industry.

2. If the performance criteria test were properly developed and the course materal structured to produce a terminal behavior to meet the performance criteria, the student would not only be certificated by the FAA but also be given collegiate credit which could be applied toward a degree, if the student chose to further his or her education at a later date.

3. The performance criteria test approach would lead to a standardization of curricula for training maintenance technicians in all schools, whether junior colleges, or universities (there are over 400 regionally accredited schools offering some type of aviation involvement), as is being done in some schools with medical technicians. It would also establish a basis for awarding credits and, also a basis for the transferring of credits from one institution to another.

These types of changes, if incorporated in the FAA's regulations, would certainly enhance the workmanship of new mechanics and, accordingly make the manager's job a somewhat easier task. These are interesting recommendations.

While the records indicate that airline and general aviation accidents and fatalities have shown a marked decline in number in recent years, there appears to be major concern whether this trend will continue. The 1980s began very positively for the certificated scheduled air carriers flying large aircraft. In 1980, there were no fatal accidents involving these carriers. There was an accident in 1980 involving a supplemental (unscheduled) airline aircraft which struck and killed a parachutist in flight near San Diego, California. This trend of no fatal accidents involving scheduled air carriers continued into 1981 setting a record 20 months of flying by the scheduled airlines without a fatal accident.[6] In 1981 there were four separate fatal accidents; however, they involved a single fatality in each accident, rather than the expected multiple fatalities. In 1982 there were five fatal accidents, including the well-publicized Air Florida accident in Washington, D.C. These three accidents caused a total of 235 deaths, the third worst in the previous decade. The 1983 results were better with only three fatal accidents claiming 14 lives. One of the accidents was the ill-fated Air Illinois Flight 710 crash near Pinckneyville, Illinois, in October 1983, which claimed 10 lives.

Commuter airline fatalities have declined in each successive year in the 1980s. The number of total fatal accidents is also down overall, with the number of accidents dropping from 15 in 1979 to two in 1983. The numbers for 1984 are not published yet, but there is expected to be a slight upturn in these figures for the year 1984. General aviation and on-demand air taxis achieved substantial reductions by 1982 in both total and fatal accident rates—as much as 17 percent in the air taxi fatal accident rate. General aviation accident totals were 10-year lows. Despite the upturn in the fatality toll and fatal accident rate of the airlines, accident statistics in all categories of civil aviation generally continued long-term downward trends. The overall accident rates of airlines in 1973 were more than twice their rates in 1982.

Two single-fatality accidents, in addition to the crashes in Washington, Boston, and New Orleans, gave the airlines (carriers operating large aircraft) five fatal accidents in 1982. There was only a series of four bizarre single-fatality accidents in 1981. Those four accidents produced a distorted rate of 0.061 fatal accidents in every 100,000 aircraft *hours* flown in 1981. Fatal accidents in 1982 resulted in a similar rate of 0.062. (See Table 2, for rate per 100,000 aircraft *departures*). The airlines' total accident rate dropped from 0.381 to 0.232 per 100,000 hours—down 39 percent. There were 16 accidents of all kinds, as compared to 25 in 1981. The rate of 0.232 was only 5 percent above the record low of 0.221 recorded in 1980.

Commuter air carriers had their safest year in the eight years for which their accident statistics are available.[7] Their total accidents dropped from 33 to 21; their fatal accidents from 10 to 4. As a result, total and fatal accident rates were 1.12 and 0.21 per 100,000 departures—the rate most often used to measure safety in the commuters' short-haul operations. These were respective 38 and 61 percent reductions from commuters' 1981 rates. Total commuter fatalities were down 64 percent, from 36 in 1981 to 13 in 1982. It was the fourth successive year of decrease in the commuters' total accident rate.

On-demand air taxis had 145 total accidents—the second lowest total in the eight years recorded. Thirty-two of these were fatal. The fatality toll of 75

TABLE 2
UNITED STATES AIR CARRIERS' SCHEDULED SERVICE AND ACCIDENT STATISTICS, 1973–1983

	Departures (millions)	Fatal Accidents	Fatalities	Fatal Accidents Per 100,000 Departures
1973	5.1	8	221	0.156
1974	4.7	7	460	0.127
1975	4.7	2	122	0.043
1976	4.8	2	38	0.041
1977	4.9	3	78	0.061
1978	5.0	5	160	0.100
1979	5.4	4	351	0.074
1980	5.3	0	0	0.000
1981	5.2	4	4	0.077
1982	4.9	5	235	0.080
1983*	4.9	3	14	0.061

Sources: National Transportation Safety Board, *Annual Report to Congress, 1983* (Washington, D.C.: NTSB, June 1984), pp. 9, 48; *Air Transport, 1984: The Annual Report of the U.S. Scheduled Airline Industry* (Washington, D.C.: Air Transport Association of America, June 1984), p. 6.
*Preliminary figures.

persons similarly was the second lowest. The lower accident totals produced a total accident rate of 5.09 per 100,000 aircraft hours, down 5 percent from 1981, and a fatal accident rate of 1.12 per 100,000 aircraft hours, down 17 percent.

General aviation—all other civil flying—recorded lower accident totals and rates across the board. There were 3,276 accidents. Of these, 574 were fatal ones, down 4 percent from 1981. The fatal accident rate was 1.59 per 100,000 hours, an 11 percent reduction. Like the large airlines, general aviation has seen a significant and consistent downward trend in its accident rates. The reduction in total accident rate in the 10 years from 1973 to last year was 40 percent. The fatal accident rate reduction for the same period was 37 percent.

A General Aviation Manufacturers Association (GAMA) manpower report predicts serious shortages of highly trained A&P mechanics by the end of the 1980s. GAMA's report, completed after six months of revising 1981 research findings, indicates that by the end of the 1980s there will be an insufficient number of competent technical personnel to keep aircraft flying safely. The study findings revealed that in six nonpilot categories, manpower shortages currently exist and that trained, skilled technicians and mechanics will be in very short supply for the aviation industry by the end of the decade. The study highlights critical mechanic needs. This points to a mandatory increase in surveillance by the maintenance manager of all completed, maintenance work orders and to an increased responsibility in screening new hires.

Man's first powered flight covered 120 feet.[8] The progress in air transportation since that windy day of December 17, 1903, at Kitty Hawk, North Carolina, can be summed up in this single comparison: You can easily duplicate the entire length of that historic flight inside the cabin of a modern jetliner and still have room to spare. "Aviation's achievements are impressive enough, but they are even more remarkable when silhouetted against the backdrop of their brief time-span—three-fourths of a century is but a mere heartbeat in history. . . . Orville and Wilbur Wright launched not just an airplane, but a technological breakthrough." The technological breakthrough might be grasped more fully by comparing the Wright Flyer with modern jetliners, Table 3. United States air transportation is a system, literally linking the smallest town with the rest of the nation and, for that matter, the rest of the world.

The airlines in 1979 accounted for more than 80 percent of all public intercity passenger miles, carried nine out of every ten intercity first class letters, served more than 400 airports, and carried more than five billion ton miles of freight annually, much of it priority material demanding fast delivery. Entire industries have been changed radically by the commercial air-

TABLE 3
DEVELOPMENT OF UNITED STATES AIR TRANSPORTATION

	Wright Flyer	Modern Jetliners
Speed	31 MPH	500–600 MPH
Range	¼ Mile	1,000–6,000 Miles
Horsepower	16	10,000–75,000 Each Engine
Passengers	1	80–400
Length	21 Feet	100–230 Feet
Wingspan	40 Feet	90–195 Feet
Weight (empty)	650 Lbs.	50,000–360,000 Lbs.

Source: Robert J. Serling, *Wrights to Wide-Bodies: The First Seventy-Five Years* (Washington, D.C.: Air Transport Association of America, 1978).

plane—food, to mention one. Lettuce picked in California on a Monday can be eaten in a New England home Tuesday night. Lifesaving medicine, emergency replacement parts to prevent major industrial outages—even valuable horses—these are some of the thousands of items dependent on air transportation. And so are people. Each of more than 400 United States airports handles more flights a year than London's Heathrow Airport, Europe's busiest. Table 4 shows the number of passengers enplaning at the 30 busiest United States airports in 1983. The technological miracle that is the jetliner has altered the nation's travel habits. Pittsburgh is next door to Paris, Binghamton around the corner from Bombay, Tulsa only a half-day from Tokyo.

The number of types of aircraft, both scheduled airline and general aviation, will vary by the end of any given month or year. Recent figures are shown in Tables 5 and 6.

The scope of aviation activity continues to appear to be almost unlimited, as powered heavier-than-air flights has passed its eightieth anniversary. Looking back over its comparatively short history, one cannot help realizing that the growth of aviation has been an era of superb research

TABLE 4
PASSENGER ENPLANEMENTS AT THE 30 BUSIEST UNITED STATES AIRPORTS IN 1983

Chicago O'Hare	19,116,000	Minneapolis	5,909,000
Atlanta	18,811,000	Pittsburgh	5,644,000
Los Angeles Int'l	15,991,000	Seattle-Tacoma	5,272,000
New York Kennedy	13,240,000	Detroit	5,075,000
Dallas-Fort Worth	12,861,000	Las Vegas	4,809,000
Denver	11,936,000	Phoenix	4,675,000
San Francisco Int'l	10,364,000	Philadelphia	4,544,000
Miami	9,153,000	Tampa	3,838,000
New York LaGuardia	9,076,000	Orlando	3,767,000
Boston	8,617,000	Charlotte	3,572,000
St. Louis Int'l	7,626,000	Salt Lake City	3,318,000
Newark	7,584,000	San Diego	3,113,000
Honolulu	7,193,000	New Orleans	3,063,000
Washington National	6,805,000	Dallas Love Field	2,930,000
Houston Intercontinental	6,402,000	Cleveland	2,745,000

Source: Federal Aviation Administration, *FAA Terminal Area Forecasts FY 1985–1995* (Washington, D.C.: USDOT/FAA, 1984).

TABLE 5
AIRCRAFT IN SERVICE (UNITED STATES SCHEDULED AIRLINES)

Manufacturer	Model	1972	1981	1982	1983*
Airbus Industry	A300		24	30	34
Beech	BE-90/99		34	23	
Boeing	B-707	337	129	54	
	B-720	56		1	
	B-727	662	1,042	965	1,038
	B-737	134	249	289	283
	B-747	106	148	141	148
	B-757			2	15
	B-767			17	55
British Aircraft Corp.	BAC-1-11	58	27	35	24
Convair	580/600/640	135	57	56	15
	880	41			
de Havilland	DHC-6/7		29	48	11
Fairchild	F-27/227	61	1	13	
Lockheed	L-188/328B/100	22	61	57	2
	L-1011	17	104	111	136
McDonnell Douglas	DC-8	227	139	131	102
	DC-9	329	444	498	555
	DC-10	59	155	162	154
Nihon	YS-11	22	9	3	
Swearingen	Metro		47	59	
Other		95	109	135	147
Total		2,361	2,808	2,830	2,719

Sources: *Air Transport, 1983: The Annual Report of the U.S. Scheduled Airline Industry* (Washington, D.C.: Air Transport Association of America, June 1983), p. 8; *Air Transport, 1984: The Annual Report of the U.S. Scheduled Airline Industry* (Washington, D.C.: Air Transport Association of America, June 1984), p. 9.
*ATA members only.

and exceptional achievement.[9] "It took centuries for man to progress from walking on foot at three to six miles an hour or riding animals at 35 to 45 miles an hour, to driving steam locomotives and automobiles at well over 100 MPH. Yet it took only about half a century of aviation history to exceed the speed of sound."[10]

TABLE 6
UNITED STATES AIRLINE AND GENERAL AVIATION FLEETS

	1973	1982	1983
U.S. air carriers*			
Total aircraft	2,614	3,838	3,585
Turbine	2,463	3,306	3,043
Piston	138	526	530
Rotorcraft	13	6	12
% of total aircraft	1.7	1.8	1.7
General aviation			
Total aircraft	153,540	209,799	207,000†
Turbine	3,271	9,182	9,700
Piston	144,875	189,195	185,300
Rotorcraft	3,143	6,160	6,700
Other	2,251	5,233	5,300
% of total aircraft	98.3	98.2	98.3

Source: Air Transport, 1984: The Annual Report of the U.S. Scheduled Airline Industry (Washington, D.C.: Air Transport Association of America, June 1984), p. 8.
*Includes scheduled, supplemental, commuter, air taxi, and cargo carriers.
†Preliminary.

2

The Federal Aviation Administration

NOW AN OPERATING ARM OF THE Department of Transportation, the Federal Aviation Administration[1] traces its history back to the Air Commerce Act of 1926, which led to the establishment of the Aeronautices Branch (later reorganized as the Bureau of Air Commerce) in the Department of Commerce with authority to certificate pilots and aircraft, develop air navigation facilities, promote flying safety, and issue flight information.

The federal government acted just in time. In May 1927, Charles Lindbergh bridged the North Atlantic in thirty-three and one-half hours, generating new interest and enthusiasm for aviation in both Europe and America. Aviation continued to grow and expand at a very rapid rate in the decade following Lindbergh's flight, creating a need for new machinery to regulate civil flying. The result was the Civil Aeronautics Authority with responsibilities in both the safety and economic areas. In 1940, the machinery was readjusted and the powers previously vested in the Civil Aeronautics Board (CAB), was placed under an Assistant Secretary in the Department of Commerce, and a semi-independent Civil Aeronautics Board (CAB), which had administrative ties with the Department of Commerce but reported directly to Congress.

In 1958, the same year American jets entered commercial service, Congress passed the Federal Aviation Act, which created the independent Federal

Aviation Agency with broad new authority to regulate civil aviation and provide for the safe and efficient utilization of the nation's airspace.

In April 1967, the Federal Aviation Agency became the Federal Aviation Administration and was incorporated into the new Department of Transportation, which had been established to give unity and direction to a coordinated national transportation system. The FAA's basic responsibilities remained unchanged, however. While working with other administrations in the Department of Transportation in long-range transportation planning, the FAA continues to concern itself primarily with the promotion and regulation of civil aviation to ensure safe and orderly growth. This effort concerns itself with the following areas of responsibility: air traffic control, aircraft and airmen certification, airport aid and certification, protecting the environment, the Civil Aviation Security Program, engineering and development, and other major activities.

One of the FAA's principal responsibilities is the operation and maintenance of the world's largest and most advanced air traffic control and air navigation system.[2] Almost half the agency's work force of 55,000-plus people are engaged in some phase of air traffic control. They staff some 400 airport control towers, 25 air route traffic control centers, and over 300 flight service stations.

An additional 10,000 technicians and engineers are required to install and maintain the various components of this system such as radars, communications sites, and ground navigation aids. For example, the system includes more than 275 long-range and terminal radar systems, more than 600 instrument landing systems, and approximately 950 very high frequency omnidirectional radio ranges. The FAA operates its own fleet of specially equipped aircraft to check the accuracy of this equipment from the air.

Almost all airline flights—and many general aviation (nonairline) flights—operate under instrument flight rules (IFR) regardless of weather conditions. This means they are followed from takeoff to touchdown by air traffic control to ensure that each flies in its own reserved block of airspace, safely separated from all other air traffic in the system.

A typical transcontinental flight from Los Angeles to New York, for example, involves almost a dozen air traffic control facilities. From the tower cab at Los Angeles International Airport, the flight is transferred, or "handed off," first to the terminal radar control room and then to the air route traffic control center at Palmdale, California. The Salt Lake City center takes control next and, depending on the route, may be followed by the Denver, Kansas City, Chicago, Cleveland, and New York centers. Approximately 30 miles from John F. Kennedy International Airport, the flight is handed off to the radar approach control facility serving all New York airports and, finally, to the JFK tower cab, which issues final landing instructions. Only when the aircraft is safely on the ground and has taxied clear of other traffic does the FAA's responsibility and concern for the safety of the passengers and crew on that particular flight end.

In order to keep pace with the rapid growth of aviation, the FAA has implemented a computer-based semiautomated air traffic control system at all of the 20 enroute centers serving the contiguous United States and at all major terminal facilities. The system tracks controlled flights automatically and tags each target with a small block of information written electronically on the radar scopes used by controllers. Included in this data block are aircraft identity and altitude—information which previously had to be acquired by voice communications, thereby imposing a burden on both pilots and controllers and contributing to radio frequency congestion and providing the possibility of human error.

Similar automated radar systems, tailored to the varied traffic demands of terminal locations, already have been installed and are operational at more than 60 large and medium hub airports. Another 80 systems have been installed at airports in the small hub category. FAA plans call for the enroute and terminal systems to be tied together nationwide in a common network for the exchange of data. The capabilities of the automated system also are being upgraded to include additional air traffic management functions such as automatic prediction and resolution of air traffic conflicts, metering and spacing of enroute aircraft, and flow control of aircraft in the terminal area.

No air traffic control system, no matter how automated, can function safely and efficiently unless the people and machines who use the system measure up to certain prescribed standards. The FAA, therefore, has been charged with responsibility for establishing and enforcing standards relevant to the training and testing of airmen and the manufacture and continued airworthiness of aircraft. There are over 217,000 (December 1982) civil aircraft in the United States, and the FAA requires that each be certificated as airworthy by the agency. Both the original design and each subsequent aircraft constructed from that design must be approved by FAA inspectors. Even home-built aircraft require FAA certification.

In the case of large transport aircraft, such as the wide-body jets, the certification process may take a number of years (figure 1)[3] The FAA's involvement begins when the aircraft is still in the blueprint stage. FAA aeronautical engineers work side by side with factory engineers throughout the entire developmental, prototype, and production process, checking on the progress of the numerous components such as the fuselage, wings, landing gear, and tail surfaces to assure quality of workmanship and conformity to an approved design. The same watchfulness is exercised over the design and manufacture of aircraft engines, propellers, and instruments.

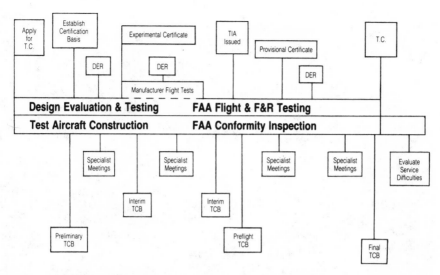

Figure 1. Type certification process (Courtesy of The Boeing Company)

When the new aircraft prototype is finished, it must pass an extensive series of ground and flight tests. If all goes well, the airplane receives a *Type Certificate* to show that it meets FAA standards of construction and performance. This is followed by the issuance of a *Production Certificate* to the manufacturer when its capability to duplicate the type design has been established. Finally, each airplane off the line receives an *Airworthiness Certificate* attesting to the fact that it conforms to the Type Certificate and is safe to fly.

Small aircraft (see Figure 2) get the same close attention during design, construction, testing, and production as do the big ones. Some factories do a sufficient volume of business to require FAA inspectors at the job full-time; others may not, but the procedures are identical and FAA inspectors personally make final checks. Service-related problems must be reported to the FAA. For this purpose, the FAA utilizes, extensively, General Aviation Airworthiness Alerts which are, for the most part, prepared from information supplied by those who operate and maintain aircraft.

GA Airworthiness Alerts provide a common communication channel through which the aviation community can economically interchange service experience and thereby cooperate in the improvement of aeronautical product durability, quality, reliability, and safety. The contents incude items that have been reported to be significant, but which have not been evaluated

Figure 2. 1983 Piper Dakota (Courtesy of Piper Aircraft Corporation)

fully by the time the material went to press. As additional facts, such as cause and corrective action, are identified, the data is published in subsequent issues of the Alerts. This procedure gives Alerts readers prompt notice of conditions reported via Malfunction or Defect Reports (FAA Form 8010-4).

To assure good maintenance practices and the adherence to all FAA requirements, aircraft manufacturers include in their sales brochures the publications available to operators and owners. In general aviation aircraft, for example, Beech Aircraft Corporation makes available the following (for each Beechcraft model):

1. Aerofiche copies
2. Paper copies
   a. Pilot's Manuals
      FAA Flight Manuals
      Owner's–Operator's Manuals
      Pilot's Handbooks
   b. Shop/Maintenance Manuals
      Overhaul Manuals
      Line Service Manuals
   c. Wiring Diagram Manuals

d. Parts Catalogs
e. Miscellaneous
   Check Lists
   Power Charts
   Inspection Forms[4]

In addition to the above listed manuals, Beech Aircraft has established authorized Beechcraft service on all six continents—Beechcraft Jet Centers, Aviation Centers, and Aero Centers. Factory-trained technicians backed by Beechcraft factory service engineers are on call 24 hours a day, seven days a week, to provide service support whenever necessary.

Once an aircraft starts flying, the FAA is concerned that it remain airworthy. Therefore, the FAA approves airline maintenance programs, setting the times for periodic inspections and overhauls of various aircraft components such as engines, propellers, instruments, communications, and flight system. The FAA also certificates, or licenses, repair stations that perform the required maintenance checks as well as repairs and alterations on general aviation aircraft (those flown by business persons, by commercial and industrial operators, air taxi operators, and private owners). All of these facilities are checked at regular intervals by FAA inspectors.

The end result of all these efforts is reflected in statistics which show that mechanical or structural defects account for only a relatively small percentage of aviation accidents. The key element in the safety equation is still the human one. For this reason, the FAA requires that everyone directly involved in the operation, maintenance, and direction of airplanes has a valid certificate from the agency with appropriate ratings. Included are pilots, flight engineers, navigators, aviation mechanics, air traffic controllers, aircraft dispatchers, and parachute riggers. In addition, the FAA certificates both pilot and mechanic schools and the instructors who teach in these institutions. There are presently almost 800,000 certificated pilots and 350,000 other airmen.

One of the FAA's most significant efforts is aimed at expanding and modernizing the nation's airport facilities to meet projected traffic demands through the 1980s. The agency was given broad new power to pursue this objective by the Airport and Airway Development Act of 1970, which replaced the Federal Airport Act of 1946 and established the Airport Development Aid Program (ADAP) and the Planning Grant Program (PGP). (By act of Congress, 1982, ADAP has been changed to the Airport Improvement Program.)

Under the Airport Improvement Program, the FAA is authorized to allocate funds for airport improvement and contruction projects. And, during the first nine years of the new program, the agency allocated more development money than it did during the entire 26-year history of the previous Federal-Aid Airport Program (FAAP). Funds are allocated on a cost-sharing basis for such projects as acquisition of land, construction of runways, taxiways, and aprons, purchase of fire/crash/rescue equipment, and installation of lighting and navigation and landing aids, and even new or enlarged passenger vehicle parking lots.

The purpose of the Planning Grant Program is to promote the orderly and timely development of the nation's airport system by assisting state and local authorities in identifying present and future air transportation requirements. Grants are made for two types of planning projects: (1) preparation of master plans at individual airports and (2) development of statewide or regional airport system plans. The FAA pays three-fourths of the cost of a planning project with the local agency contributing the remainder.

The Airport and Airway Development Act of 1970 also authorized the FAA to issue operating certificates to airports receiving service to assure their safe operation. In keeping with this directive, the agency subsequently adopted new regulations setting safety standards in some 18 areas including the availability of fire-fighting and rescue equipment, reduction of bird hazards, marking and lighting of runways and taxiways, handling and storage of dangerous materials, and marking and lighting of obstructions.

The first phase of the certification program was limited to the approximately 500 airports that receive regular scheduled service by Civil Aeronautics Board-certified air carriers using large aircraft and account for 96 percent of all airline passenger enplanements in the United States. The agency completed certification of these airports in May 1973. The second phase involved those airports serving CAB-certificated air carriers conducting operations on an irregular or unscheduled basis or operations with small aircraft. FAA operating certificates had been issued to more than 700 airports by mid-1979. The FAA also assists airport owners in designing, constructing, and maintaining airports in keeping with aviation requirements, national standards of safety, and efficiency in terms of the state-of-the-art in design and engineering technology. This is accomplished by the issuance of standards, published in the form of Advisory Circulars, which are mandatory for grant recipients and have worldwide acceptance as technical advisory documents. Advisory Circulars in this respect cover such areas as airport paving, airport drainage, runway/taxiway/apron design, and airport lighting.

In addition to safety, the FAA also has the important responsibility of making airplanes compatible with the environment by controlling noise and engine emission characteristics. The agency considers these efforts of critical importance in ensuring the future growth and development of civil aviation in the United States.

A significant measure of the progress already made in this area is the performance of the newer generation of wide-body jets, such as the Boeing 747, DC-10, and the Lockheed 1011. Although the engines which power these aircraft generate two and one-half times the thrust of any engine previously used in commercial service, they are only about half as loud (or twice as quiet) as their predecessors. In addition, they are virtually smoke-free.

The FAA also has initiated regulatory action designed to quiet older jets presently in service by requiring that they either be modified with noise suppression devices or be phased out of service. In addition, engine noise standards have been developed for the new generation of aircraft and the supersonic transports. The FAA already has adopted a regulation which prohibits flights by civil supersonic aircraft over the United States if they would result in a sonic boom reaching the ground.

A similar effort to clean up aircraft engines also is underway with FAA acting in concert with the Environmental Protection Agency. Under the Clean Air Act of 1970, the EPA is responsible for setting standards governing the emissions of noise, carbon monoxide, hydrocarbons, and nitrogen oxides for virtually all engines used in civil aircraft. The FAA, in turn, serves in an advisory capacity in the establishment of these standards and has final authority for their implementation through regulatory action.

Another major FAA responsibility is the Civil Aviation Security Program. Efforts in this area are aimed at preventing or deterring such criminal acts as air piracy, sabotage, extortion, and other crimes which could adversely effect aviation safety. Key elements of the program include a requirement for the screening of all enplaning airline passengers and a search of their carry-on baggage. A law enforcement officer also must be present at each screening station during the boarding process. In addition, airport operators are required to establish a security system that will keep unauthorized persons from gaining access to air operations areas.

Implementation of these regulations early in 1973 and the negotiations of an agreement with the Cuban Government on the disposition of hijackers at about the same time produced a dramatic turnaround in the hijacking situation. After averaging almost 30 per year during the 1968–72 period, the number of hijacking attempts dropped to 5 in 1977 and 8 in 1978, none of which was successful.

Civil aviation security was strengthened further in August 1974, when Congress passed the Anti-hijacking Act of 1974, which gave statutory force to the FAA's security regulations. And, in July of 1978, the industrialized nations of the world agreed at a summit meeting in Bonn, West Germany, to act together to cut off all air service to and from countries that refuse to extradite or prosecute aircraft hijackers.

The FAA supports all of its safety, security, and environmental programs with extensive engineering and development (E&D) projects, conducted in part through contracts with industry, other government agencies, and universities. Much of the E&D work however, is done in-house at the FAA's Technical Center at Atlantic City, New Jersey, and the Transportation Systems Center at Cambridge, Massachusetts. Aeromedical research is done at the FAA's Civil Aeromedical Institute at Oklahoma City.

A continuing priority of the agency's E&D effort is further automation of the air traffic control system to help controllers keep aircraft safely separated as air traffic increases. Warning systems, for instance, have been added to the automated systems at the busiest air traffic facilities to alert controllers when aircraft under their control are dangerously close to the ground or too close together. Work is under way to develop other computer add-ons that will assist controllers in handling higher traffic loads with increased efficiency and safety.

The FAA also is developing collision avoidance systems that will operate independently of the air traffic control system but will be compatible with it. These electronic devices will warn pilots directly of potential conflicts with other aircraft and show how to avoid them. The first of these systems, designed for use in en route airspace and at airports with light to moderate traffic, was ready in late 1981. In the meantime, the FAA will continue developing and testing more sophisticated collision avoidance systems for effective operation in congested airspace.

An important element in an effective collision avoidance system for high use airspace is the Discrete Address Beacon System (DABS) which is being developed by the FAA to upgrade the present air traffic control radar beacon surveillance system. Essentially, DABS is an improved transponder but it also will provide a data link for use with a ground-based anticollision system, and it will be the basis for other system improvements such as automatic metering and spacing to improve the flow of traffic and automatic weather reporting.

Supplying pilots with accurate and timely weather information, particularly about hazardous weather, is another major E&D program goal. Among the efforts under way to achieve this safety goal are the development and demonstration of automated weather observation systems for airports without control towers, testing of a wake vortex advisory system that warns pilots of potentially dangerous air turbulence in approach and departure paths, and low level wind shear alert systems to help pilots cope safely with wind shear during the critical stages of approach and landing. In addition to enhancing safety, these weather systems will help reduce delays, conserve fuel, and allow more efficient use of airport capacity.

Another major E&D effort nearing completion is development of a Microwave Landing System (MLS) to replace the present Instrument Landing System at airports around the world. Initial operational use began in 1980. MLS is being installed at airports where ILS cannot be used because of mountainous terrain and other siting problems. It also provides much greater flexibility in the use of terminal airspace by permitting curved approaches and a more varied choice of approach angles for all types of aircraft. The MLS service is far more accurate; it can discern altitude errors of as little as 2 feet and horizontal errors of 13 feet or less when an aircraft crosses the runway threshold. Landings are possible with less visibility and with lower ceilings. The major advantage with the MLS system is the broad funnel-shaped signal it sends out. Since pilots can fly approaches from many different angles, this eliminates the need to come from a straight, narrow approach, allowing the controllers to be able to get planes on the ground more quickly. The MLS can also guide aircraft in through automatic landings. The end result is reduced congestion and delay at airports along with noise relief in surrounding communities.

The FAA also has an extensive aeromedical research program to explore the human factors which affect the safety and advancement of civil aviation. Some of the research efforts under way include studies of crash impact and survival, the toxic hazards of burning cabin materials following a crash, and the effect of aging and stress on airman performance.

Because the United States is the recognized world leader in aviation, the FAA has a vital role to play in international aviation matters. For example, in cooperation with the State Department's Agency for International Development, it sends Civil Aviation Assistance Groups abroad to provide technical aid to other nations and also trains hundreds of foreign nationals every year at the FAA's Mike Monroney Aeronautical Center in Oklahoma City.

It also works with the International Civil Aviation Organization (ICAO) in establishing worldwide safety and security standards and procedures, provides technical advice on the export and import of aviation products, and handles certification of foreign-made aircraft engines and parts under the terms of bilateral airworthiness agreements.

The FAA also participates with the National Transportation Safety Board (NTSB) in the investigation of major aircraft accidents to determine if any immediate action is needed to correct deficiencies and prevent a recurrence. In addition, the agency investigates most nonfatal and many fatal general aviation accidents on behalf of the NTSB, although the responsibility for determining probable cause remains with the Board.

Finally, the FAA operates the two airports serving the nation's capital—Washington National Airport, located just 4 miles from the heart of the city, and Dulles International Airport, 26 miles west of Washington in Chantilly, Virginia. The FAA also operates a public-use airport at its Technical Center outside of Atlantic City, New Jersey.

For the reader's interest, The "FAA Organization" (manual) contains the following orders.[5]

Order 1100.1A, FAA Organization—Policies and Standards, contains general organizational policies, standards, concepts, and philosophy applicable throughout the FAA.

Order 1100.2A, Organization—FAA Headquarters, prescribes the organization and functions of the offices and services to the division level.

Order 1100.5A, FAA Organization—Field, prescribes the organizational structure and functions to the division level for the regions, the Aeronautical Center, the FAA Technical Center, Metropolitan Washington Airports, and the Europe, Africa, and Middle East Office, but includes elements below this level when prescribed by the Administrtor, i.e., prescribed branches in the regional Air Traffic Division.

Order 1100.14A, FAA Organization Manual, contains the mission and functional statements for the FAA organizational elements approved by the Secretary of Transportation.

The "FAA Organization" (manual) is the principal medium by which the Administrator establishes major organizational concepts and structure, assigns missions and functions, and delegates authority.

On these matters, the manual is the basic authority within the agency. All agency activities and documents which reflect organization, assignment of responsibility, or delegation of authority must conform to the manual. This includes, but is not limited to, documents such as supplementary organization directives, position descriptions, telephone listings, and information released to the public. The manual may be overridden only after written authority is obtained from the Administrator. These decisions by the Administrator shall be followed by amendments to the manual.

The manual covers FAA organization generally down to and including the division level in offices, services, regions, centers, the Europe, Africa, and Middle East Office, and the Metropolitan Washington Airports, but includes elements below these levels when prescribed by the Administrator.

Organizational material for organizations established below the levels prescribed by the manual shall be published in supplementary organizational directives issued by the appropriate region, office, service, the Aeronautical Center, FAA Technical Center, the Europe, Africa, and Middle East Office, or Metropolitan Washington Airports. These directives have the same standing and authority within their areas of coverage as the manual has for the agency.

The new FAA Regional Boundaries are seen in Figure 3. The FAA airworthiness technical staff and field functions are also outlined. It should be recognized that the standard regional organization concept has been replaced by the basic regional organization concept,[6] effective July 1, 1981.

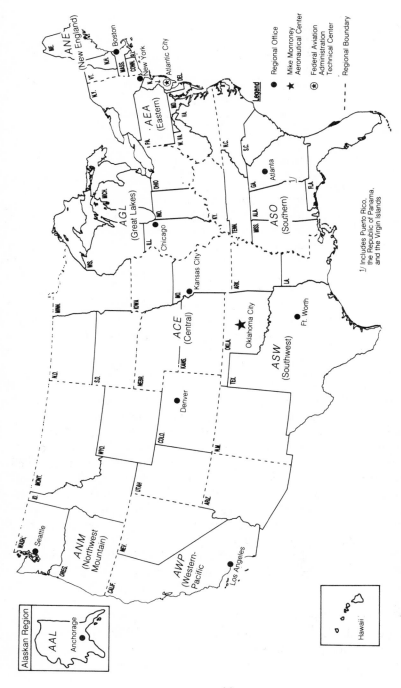

Figure 3. FAA regional boundaries (Order 1100.5A, Change 78, May 17, 1982)

FAA AIRWORTHINESS TECHNICAL STAFF

    Aerospace Engineers
    Flight Text Pilots
    Aviation Safety Inspectors
        —Manufacturing
        —Airworthiness (Maintenance)
        —Avionics (Maintenance)

FAA AIRWORTHINESS FIELD FUNCTIONS

    Engineering and Manufacturing—
        Centralized in Regional Office with Support from EMDOs (Engineering and Manufacturing District Offices)
    Maintenance with Operations—
        Centralized in District Offices:
            —ACDOs (Air Carrier District Offices)
            —GADOs (General Aviation District Offices)
            —FSDOs (Flight Standard District Offices)

The regulatory basis for approval of air carrier maintenance programs is contained in Federal Aviation Regulations, Part 121.[7] The Federal Aviation Administration is responsible for the approval of air carrier maintenance programs (the programs that establish the time limitations or standards for determining time limitations for overhauls, inspections, and checks of airplanes, engines, and appliances). The characteristics of these limitations and standards have been subject to continued change as changes in the state-of-the-art of airframe, engine, and appliance design have occurred and knowledge about the effectiveness of preventive maintenance has been obtained.

The oldest recognized primary maintenance process, generally called "hard-time," requires periodic overhaul or replacement of the affected hardware. During the early days of commercial aviation, "hard-time" was generally considered to be the most effective maintenance process. It was applied with the intent of ensuring operating safety of airplanes having limited systems redundancy.

After World War II, the FAA recognized that, for some hardware, checking to a physical standard at periodic intervals was also an important and effective maintenance process. This process, called "on-condition," was the second primary maintenance process to be recognized. At that time, "hard-time" and "on-condition" were the only recognized primary maintenance processes. Because there were only two alternatives, gradually "on-condition" was applied to many items where neither alternative was appropriate.

In the 1960s, the FAA released Advisory Circular 120-17 and approved a number of "reliability programs." These permitted air carriers to explore the relationship between age and reliability without conventional time limitations. A wide range of programs have been approved. Some use "hard-time" limitations; some use "on-condition" physical standards; and some use only reliability performance standards to manage reliability. Experience with programs using only reliability performance standards made it clear that some aircraft elements did not require the traditional preventive primary maintenance process in order to ensure operating safety.

From this experience came the development by the industry 747 Steering Group and an FS-300/747 Advisory Group of a new technique for the design of initial maintenance programs. This technique, which requires intensive review of the aircraft design by industry and FAA specialists and application of a process called "decision tree analysis," is currently used for all new air transport initial maintenance programs. Historically, it came about as follows:

For two years, an industry team explored a maintenance concept that reflected the special concerns of the time. The study team wanted a planning document that would address not only the new technology of the 767, but also some of the more significant industry concerns.

What the study team produced was MSG-3 (short for Maintenance Steering Group, Revision No. 3). As "Revision No. 3" suggests, this new planning document evolved from prior studies and built on maintenance practices established for earlier generation wide-bodies. (MSG-1 was developed for the Boeing 747, while MSG-2 was adopted for the DC-10 and L-1011.)

MSG-3 addresses such issues as the FAA's damage tolerance requirements for aircraft structures (FAR 25.571), the Lowe Committee findings on FAA-carrier relationships, the maintenance of aging aircraft, and the NTSB findings on the DC-10 tragedy at O'Hare (May 1979). The economic guidelines in MSG-3 even show the influence of skyrocketing fuel costs on maintenance decisions. The study team did its homework well, and MSG-3 was given approval by the FAA Chief of Airworthiness for use on the 767.

As with its predecessor programs, MSG-3 is based on a consistent and rigorous application of questions for each collection of aircraft hardware. It is decision tree analysis at work. The first question MSG-3 asks is: "What's the consequence of a specific hardware failure on the entire aircraft?" Once this consequence is assessed, MSG-3 offers a choice of applicable tasks and evaluates each one's effectiveness. Once a task is chosen, its frequency is patterned after frequencies adopted for similar hardware. If no comparison can be made, a conservative frequency is initally adopted and adjusted as experience is gained.

An initial concern was that hardware failures directly related to airworthiness might be subject to the exploration techniques of MSG-3. This will not be the case. Instead, critical hardware is assigned discard limits to forestall service failure. (Blade-supporting discs in power plants and landing gear structures are about the only items remaining in this category.)

Returning to the consequences of a specific hardware failure, five questions form the basis for exploration.

1. Will the failure cause a direct, adverse effect on operating safety? A failure like this could cause an out-of-pilot-control condition which can endanger passenger, crew or aircraft.

2. Will the failure lead to loss of hidden function backup, which could (if coupled with a related primary system failure) create a potentially adverse effect on operating safety? Examples in this category are fire extinguisher failure or malfunction of the duct over-temperature sensing device.

3. Will the failure cause an adverse economic effect on airline operation? Failure in this category can result in turnbacks, diversions, altitude restrictions, and dispatch prohibitions.

4. Will the failure cause the loss of hidden functions, which impose an economic penalty by themselves or in conjunction with the failure of some other primary system? One failure in this category is the stuck "on" vortex-dissipating jet pump on some engines.

5. Will the failure cause an economic effect on the airline in nonoperational ways? Items in this category most often have an economic impact based on the volume of removals-for-cause.

Given these five possible consequences, each functional failure and its effect on the aircraft is evaluated and the most appropriate task is selected.

Here are the tasks that have been successfully used in the past, listed in an order of increasing economic penalty to the United Airlines: (1) service and lubrication tasks, (2) flight crew monitoring for signs of impending failure, (3) inspection (of zones and structures) or functional checks (measuring systems' resistance to failure), (4) operational check (does the system work or not), (5) restoration (remove unit from aircraft and route to home shop), and (6) remove unit from aircraft and discard.

To facilitate analysis, the aircraft's system and components are grouped in manageable "families." Failure modes, failure effects and functions are listed for each, and the analysis is made, unit by unit, until each system is analyzed.

The MSG-3 program for power plants follows that of system and components. Engine tasks, however, are geared more toward maximizing service life through restoration. Thus, the engine exploration process begins with shop visit intervals and borescope inspections which are slowly expanded until the limits of the restoration process are reached.

For airframes and structures, the program differs slightly in that failure consequences are automatically considered to have safety implications. Thus, for each major structural part, exploration begins with identification of the type, location, and size of possible structural damage. Damage sources considered are fatigue, accident, and environmental deterioration (which includes corrosion). Structural inspection ranges from daily walkaround to nondestructive tests using eddy current or X-ray techniques. Once a structural task is identified, the intervals for such tasks are selected and the analysis is complete.

This work resulted in the recognition of a third primary maintenance process called "condition-monitoring." This process applies to hardware with design characteristics warranting the use of a process not involving "hard-time" or "on-condition" checks. The outcome of this cooperative effort is reflected in the approved MRB documents.

Primary Maintenance Processes. A primary maintenance process is the process that is listed in the "overhaul period" column of the Operations Specification-Maintenance, FAA Form 1014. It is the process relied upon to ensure that inherent design reliability is maintained. The FAA recognizes three primary maintenance processes. These processes are simply a means for classifying the way in which a particular aircraft element is maintained. Any one or combination of these processes may be part of a carrier's "reliability program," developed in accordance with Advisory Circular 120-17A, or a carrier may use the conventional form of operations specifications.

The three primary maintenance processes have no self-implied order or importance. Each has its own place in an effective maintenance program. The right process is determined primarily by the design of the hardware and secondarily by the user's economics, not by any historical significance. To say it in another way, "hard-time" is not the best because it was first nor is "condition-monitoring" the best, or the worst, because it was last.

A description to each primary maintenance process follows.

1. *Overhaul Time Limit or Part Life Limit—(HT).* This is a preventive primary maintenance process (hard-time). It requires that an appliance or part be periodically overhauled in accordance with the carrier's maintenance manual or that it be removed from service. These time limitations may be adjusted based on operating experience or tests, as appropriate, in accordance with a carrier's approved reliability program, or the maintenance manual.

2. On-Condition Maintenance—(OC). This is a preventive primary maintenance process. It requires that an appliance or part be periodically inspected or checked against some appropriate physical standard to determine whether it can continue in service. The purpose of the standard is to remove the unit from service before failure during normal operation. These standards may be adjusted based on operating experience or tests, as appropriate, in accordance with a carrier's approved reliability program or maintenance manual.

3. Condition-Monitoring—(CM). This is a maintenance process for items that have neither "hard-time" nor "on-condition" maintenance as their primary maintenance process. CM is accomplished by appropriate means available to an operator for finding and solving problem areas. In effect, it obligates the user to apply knowledge gained by analysis of failures or other indications of deterioration to consider action to improve performance.

Reliability Programs. Reliability programs approved under FAR 121.25, 121.45, or 127.13 (guidelines provided in Advisory Circular 120-17A) have been adopted by many operators. These programs are essentially a set of rules and practices for managing maintenance processes. Some of these are special integrated maintenance management programs designed to meet an operator's own management needs. These may not individually recognize the three primary maintenance processes, even though they may include any or all of them and the collection of in-service operating data as well.

Continuous Surveillance and Analysis. In addition to the requirements outlined in the operator's Operations Specifications—Maintenance for specific aircraft, each operator is required by FAR 121.373 to have a system for continuous surveillance and analysis to appraise the performance and effectiveness of his or her overall program. For B-747 and later aircraft, this system will initially include all of the maintenance significant items listed in the MRB document. For earlier aircraft, the scope of this system will be determined locally, using appropriate MRB documents or Standard Maintenance Specifications. The overall surveillance and analysis system will provide procedures to ensure that all items are being maintained by appropriate primary maintenance processes.

Application of the Three Primary Maintenance Processes Concept to New Aircraft. The lack of real experience with new aircraft requires careful, detailed study of their characteristics to determine those components or systems that would probably benefit from scheduled maintenance. The initial maintenance programs for the B-747, DC-10, and L-1011 aircraft were developed by special teams of industry and FAA personnel. These teams sorted out the potential tasks and then evaluated these tasks to determine which must be done for operating safety or essential hidden function protection. The remaining potential tasks were evaluated to determine whether they were economically useful. These procedures provide a systematic review of the aircraft design so that, in the absence of real experience, the best process can be utilized for each component or system. The B-747, DC-10, and L-1011 aircraft operating experience confirmed the effectiveness of these procedures. It is expected that a similar procedure will be used for future new aircraft.

3

Regulatory Requirements

THE FEDERAL AVIATION ADMINISTRATION, the air transportation component of the Department of Transportation, regulates as well as fosters civil aviation. Its mission is directed primarily to aviation safety and to the establishment and enforcement of safety standards applicable to virtually every aspect of civil air transportation. In the discharge of its mission, the agency issues regulatory, advisory, technical, scientific, administrative, educational, and informational publications that affect civil aviation at every level.[1]

An appropriate guideline to the FAA's publications is to separate them by category: one, reflecting the FAA's regulatory and technical functions; the other, its nonregulatory and support functions.

1. *The Regulatory and Technical Functions.* The 15 categories in this group are: (1) the agency charter; (2) rulemaking materials; (3) Federal Aviation Regulations (FARs); (4) Code of Federal Regulations materials; (5) enforcement activities; (6) type certifications; (7) airworthiness directives; (8) Flight Standards forms; (9) air carrier utilization; (10) airman's information materials; (11) technical standard orders; (12) headquarters directives; (13) service difficulty reports; (14) technical and scientific reports; and (15) agency progress reports.

2. *Nonregulatory and Support Functions.* The 11 categories of publications involved here are as follows: (1) advisory circulars; (2) aeromedical reports;

Regulatory Requirements 31

(3) education and career materials; (4) agency approvals; (5) flight safety materials; (6) economic studies; (7) historical monographs and chronologies; (8) booklets, pamphlets, and brochures; (9) planning and forecast projections; (10) statistical reports and the financial report; and (11) the agency's aviation safety magazine.

The FAA regulatory and technical publications of direct concern to the maintenance supervisor and personnel are as follows:

Rulemaking Materials. Rulemaking is the process of issuing Notices of Proposed Rulemaking (NPRMs), inviting public comment on the proposed rules, holding public hearings on them, and, when a decision has been reached on how to take care of the matter, issuing new rules, directives, and requirements in the form of mandatory Federal Aviation Regulations (FARs). In an effort to ensure that its FARs are responsive to the needs and desires of the public, the agency has established a computerized procedure to make sure that those interested in commenting or testifying on NPRMs get the notices in time to take action. The rulemaking process is shown in Figure 4.

Federal Aviation Regulations (FARs). Issued by the FAA in implementation of its regulatory functions, the Federal Aviation Regulations are continuously being revised to meet changing needs. They carry the force of law and are binding on all aviation activities within their pur-

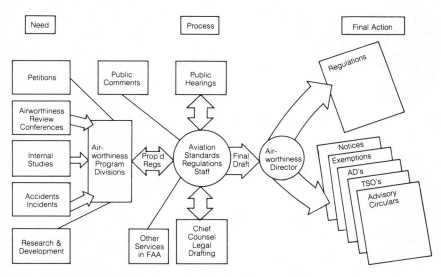

Figure 4. FAA rulemaking process (From "The FAA Organization")

view. There is a series of over a hundred regulations, each concerned with a specific aspect of aviation.

Code of Federal Regulations. A portion of Title 14 of the *Code of Federal Regulations* (CFRs) deals with FAA. It includes a codification of the general and permanent rules published in the *Federal Register* by the FAA, and brought up-to-date early each year by the office that publishes the *Federal Register.*

Title 14 CFR, as it pertains to FAA, is available in three of the four volumes of Title 14. The first is a codification of FAR Parts 1 through 59, the second of Parts 60 through 139, and the third of Parts 140 through 199.

Type Certification. The FAA issues type certificates for new aircraft models, engine models, and propeller models when it determines that they meet prescribed safety standards. In the case of an approved change in a type certificate model, the FAA issues a supplemental type certificate. If the change is substantial, it issues a new type certificate. Additional information associated with type certificates is issued in the form of type certificate data sheets and specifications. Type certificate changes are listed in a supplemental type certificate summary. The FAA has separated the roughly 1,000 data sheets in the two series into six separate volumes, based on aircraft weight and configuration, and issues in their place six separate data sheet publications. The type certificate data sheets and specifications are grouped in volumes according to aircraft weight.

Airworthiness Directives (ADs). The FAA monitors what it has certificated. An AD is issued when subsequent certified aircraft and/or parts are found to have developed unsafe conditions. ADs are discussed in greater detail in Chapter 10, "Safety and Maintenance."

Flight Standards Forms. These forms are part of the paperwork required for the implementation of the agency's airworthiness, aircraft registration, airman certification, and air carrier programs. Representative forms and the programs to which they belong are as follows:

AIRWORTHINESS PROGRAM

FAA Form 8740-5	Safety Improvement Report
FAA Form 8130-6	Application for Airworthiness Certificate
FAA Form 337	Major Repair and Alternations
FAA Form 8130-1	Application for Export Certificate of Airworthiness
FAA Form 8110-12	Application for Type Certificate, Production Certificate, or Supplemental Type Certificate
FAA Form 8600-12	FAA Inspection Reminder
FAA Form 8010-4	Malfunction or Defect Report

Airman Certification Program

FAA Form 8110-14	Statement of Qualifications (DMIR-DER-DPRE-DME)
FAA Form 8610-1	Mechanic's Application for Inspection Authorization
FAA Form 8610-2	Airman Certificate and/or Rating Application

General Aviation/Air Carrier Program

FAA Form 8000-6	Application for Air Taxi Commercial Operator (ATCO). Certificate under FAR Part 135

Air Carrier Utilization. The Flight Standards National Field Office at Oklahoma City issues a single report monthly in this area. The publication—*Aircraft Utilization and Propulsion Reliability Report*—is a statistical analysis of the efficiency of the United States air carrier fleet. The report breaks down the fleet into its component types—turbine, turboprop, reciprocating, and helicopter-type aircraft—and gives authoritative figures on the performance of each type during the month in question.

Technical Standard Orders (TSO). These are standards issued by the Administrator, setting forth in detail minimum performance standards for specified articles—materials, parts, appliances, etc.—used on civil aircraft. A portion of the material, Technical Standard Order Procedures, is incorporated into FARs, Part 21, "Certification Procedures for Products and Parts."

Service Difficulty Reports. Using reports from the aviation community, the Flight Standards National Field Office at Oklahoma City issues two different types of Service Difficulty reports. One alerts recipients to reported air carrier service difficulties; the other, to general aviation service difficulties. The reports are issued to industry affiliates and others with a demonstrated need for the service.

Scientific and Technical Reports. The FAA makes available to the public various scientific and technical reports. The FAA issues a quarterly press release listing the reports (which total several hundred each year) by date of issue, numbering of pages, and NTIS number.

Non-regulatory publications of interest to maintenance personnel include the following:

FAA Advisory Circulars. The FAA issues advisory circulars (ACs) to inform the aviation public in a systematic way of nonregulatory material of interest (see Appendix A). Unless incorporated into a regulation by reference, the contents of an advisory circular are not binding on the public.

Advisory circulars are issued in a numbered-subject system corresponding to the subject areas of the Federal Aviation Regulations (FARs) (Title 14, Code of Federal Regulations, Chapter 1, FAA).[2] Appendix A contains a listing of many ACs pertaining to aviation maintenance recommendations. The FAA publishes an advisory circular checklist (AC 00-2), issued as necessary. The checklist is free to the public and has as its principal object to keep the public informed of the agency's nonregulatory advisory materials. The advisory circular checklist entries are changed with each revised checklist.

Educational and Career Materials. The publications in this category provide educators and students with aviation education and aviation career guidance materials designed to enhance and enrich aviation study programs and to attract students to aviation careers.

Agency Approvals. The FAA periodically issues listings of FAA-approved medical examiners, parachute rigger examiners, engineering representatives, parts manufacturers, materials, parts, and appliances, aviation maintenance technician schools, pilot schools, repair stations, and parachute lofts.

For optimum safety and freedom in aviation, every airman should know the regulations pertaining to his or her individual responsibilities. In addition, every airman should be familiar with certain overlapping and related rules that require cooperative efforts of operations and maintenance to discharge properly the shared responsibility of maintaining aircraft airworthiness.

The following notes are a collective review of Federal Aviation Regulations relating to the individual and cooperative responsibilities of aircraft owners, operators, pilots, mechanics, inspectors, and repair stations for maintaining the airworthiness of general aviation and air carrier aircraft.[3]

This effort to rightly divide the FARs by collecting applicable sections from FARs 43, 65, 91, 121, 127, 135, and 145 and putting them all together accomplishes a threefold objective: (1) establishes a more knowledgeable rapport between operations and maintenance airmen and agencies, (2) prevents violations of the FARs pertaining to aircraft maintenance, and (3) prevents aircraft accidents in which lack of maintenance or improper maintenance may be a factor. (The brief comments that follow each FAR reference do not constitute a legal interpretation of the regulations; they are simply a paraphrased rendition of the primary content of that section.)

PART 1. DEFINITIONS

Person	An individual or company, corporation, etc.
Operate	Use for the purpose of air navigation.

Regulatory Requirements

Maintenance	Includes inspection, repair, etc., but not preventive maintenance.
Major Alteration	An alteration not listed in the aircraft, aircraft engine, or propeller specifications: (1) That might appreciably affect weight, balance, structural strength, performance, power-plant operation, flight characteristics, or other qualities affecting airworthiness; or, (2) that is not done according to accepted practices or cannot be done by elementary operations.
Preventive Maintenance	Simple, minor, preservation, or replacement.

Part 91. General Operating and Flight Rules

91.1	Applicability—Governs operation of aircraft.
91.29(a)	Aircraft must be airworthy to be operated.
91.29(b)	Flight to be discontinued if unairworthy condition occurs.

Part 91, Subpart C. Primary Responsibility For Airworthiness

91.161(a)	Aircraft must be maintained "within or without" the United States.
91.163(a)	Owner or operator is primarily responsible for maintaining airworthiness, including AD compliance.
91.163(b)	Prescribed maintenance and persons authorized to perform.
39.3	Operate in accordance with Airworthiness Directives (ADs).

Part 43. Maintenance, Preventive Maintenance, Rebuilding and Alteration

43.1	Applicable to all certificated aircraft except experimental aircraft that have never been issued any other kind of certificate.
43.3	Persons authorized, and work they are authorized to perform.
43.3(b)	Certificated mechanics may perform as prescribed in FAR 65.
43.3(c)	Certificated repairmen may perform as prescribed in FAR 65.

Part 65. Certification: Airmen Other Than Flight Crew Members

65.81(a)	Mechanic—General privileges and limitations. Mechanics may not perform major repair or major alteration of propellers and any repair or alteration to instruments. Must have satisfactorily performed or shown his or her ability to perform before supervising or approving for return to service.
65.81(b)	Mechanics must understand how to do a specific job before it is done.
65.83	At least 6 months in the preceding 24 months or otherwise qualified by the FAA. Recent experience required.
65.85	Airframe rating: Additional privileges, 100-hour inspection.
65.87	Power plant rating: Additional privileges, 100-hour inspection.
65.89	Certificate is to be kept where the mechanic normally works.
65.95(a)	Inspection authorization. Privileges and limitations: Inspect and approve for return to service major maintenance if done in accordance with approved data. Perform annual inspection. Perform or supervise progressive inspections.
65.95(b)	Authorization to be kept available for inspection by aircraft owner or mechanic.
65.103(a)	Repairman Certificate. Privileges and limitations: May perform or supervise the specific job for which he or she is employed and certificated.
43.3(d)	Persons working under the supervision of certificated mechanic or repairman: Supervisor personally observes to the extent necessary to ensure that work is done properly. Supervisor is readily available in person for consultation. Hundred-hour and annual inspections may not be supervised.
43.3(e)	Repair stations may perform as provided in FAR 145.

Part 145. Repair Stations

145.51	Privileges of certificate
145.51(a)	Maintain or alter items for which it is rated.
145.51(b)	Approve for return to service after maintaining and altering.

Regulatory Requirements

145.51(c)	Airframe rated station may perform 100-hour, annual, and progressive inspections and return aircraft to service.
145.51(d)	Maintain or alter articles for which rated at other places under certain conditions: May not approve for return to service any major repair or alteration unless work is done in accordance with technical data approved by the FAA.
145.53	May not maintain or alter an article if it requires special technical data, equipment, or facilities that are not available to it.
145.55	Must provide personnel, facilities, equipment equal to current standards for reissuance of certificate.
145.57(a)	Perform its operations in accordance with the standards of FAR 43.
43.3(f)&(g)	Large air carrier and commercial operators.
43.3(h)	Pilot may perform preventive maintenance on aircraft owned or operated by him, not used in air taxi service. Preventive maintenance items are listed in FAR 43, App. A.
91.165	This is the general aviation "maintenance program"—three distinct owner-operator responsibilities to ensure continual aircraft airworthiness. (1) Owner or operator shall have aircraft inspected. (2) Owner or operator shall have defects repaired between inspections. (3) Owner or operator shall insure that maintenance personnel make appropriate entries in Maintenance Records.

Aircraft Inspections

91.169(a)(1)	Annual inspection within the preceding 12 calendar months. Hundred-hour inspection may not be substituted for an annual unless performed by authorized person and recorded as an annual.
91.169(b)	Hundred-hour inspection required to carry persons for hire or give flight instruction for hire. Ten hours excess time allowed, if necessary. Any excess time must be included in next 100-hour cycle.
91.169(c)	Annual or 100-hour inspection not applicable if owner or operator complies with progressive inspections (91.171).

Part 91.171. Progressive Inspection

1. Owner/operator must submit written request to District Office.
2. Owner/operator shall provide: (a) mechanic with inspection authorization, (b) current inspection procedures manual, (c) housing and equipment necessary, and (d) current technical information for aircraft.

Part 91, Subpart D

This program became effective January 23, 1973. FAR 91.217(a) outlines the requirements of this "inspection program" and states that no person may operate a large airplane or a turbopropeller powered multiengine, unless inspected in accordance with an inspection program selected from the options below. The five options as listed in FAR 91.217(b) are: (1) continuous airworthiness program currently in use by an air carrier, (2) a program being used by an air taxi operator, (3) a program being used by an air travel club, (4) a manufacturer's inspection program, and (5) any other inspection program developed by the owner and approved by the Administrator. A&P mechanics accomplishing inspections under any option must use the owner or operator's furnished inspection guides. Approval for return to service will be made in accordance with FAR 43.9(a)(5). A&P mechanics are authorized to conduct and approve for return to service aircraft inspected in accordance with the owner or operator's program under any of the five options.

43.15(a)	Performance rules for inspections: to determine whether aircraft meets all applicable airworthiness requirements.
43.15(b)	Rotorcraft inspected in accordance with manufacturer's maintenance manual.
43.15(c)(1)	For 100-hour and annual inspections, an inspection checklist that covers the scope and detail of FAR 43, App. D, shall be used.
43.15(c)(2)	Person approving for return to service shall determine satisfactory "run up" performance.
43.16	Rotorcraft inspection and work shall be in accordance with "airworthiness limitation" section of the rotorcraft manual.
91.163(c)	Rotorcraft "airworthiness limitation" must be complied with if aircraft is to be operated (replacement time, etc.).

Other Aircraft Maintenance

43.13(a)	Maintenance performance rules (general): Use methods, techniques, and practices acceptable to

Regulatory Requirements 39

43.13(b)	FAA. Use tools, equipment and test apparatus to assure that work is done according to acceptable industry practices. Use special test equipment recommended by manufacturer or the equivalent acceptable to the FAA. The work done and materials used must be of such quality that the condition of the aircraft is equal to its original or properly altered condition.

Maintenance Records

91.173(a)(b)	Registered owner or operator shall keep the records in item 1 below until the work is repeated, superseded, or for one year. 1. Records of maintenance, alterations, required or approved inspections, as appropriate, for each aircraft, engine, propeller, rotor, and appliance must include: (a) a description of the work, (b) the date of the work, and (c) the signature and certificate number of person approving work for return to service. (The foregoing record entries must be made by the person performing the work—43.9) 2. The following records must be retained and transferred with the aircraft when sold: (a) airframe total time in service, (b) current status of life limited parts, (c) current inspection status including times since last inspection, (d) time since overhaul on items required to be overhauled, (e) current status of Airworthiness Directives including method of compliance, and (f) list of current major alterations.
91.173(c)	The owner or operator must make the maintenance records available for inspection by the FAA or National Transportation Safety Board.
91.174	Transfer of maintenance records (The records in item 2 above must be transferred to the new owner and those specified in item 1 must be transferred also, unless arrangements are made with the seller to make them available to the FAA or NTSB on request.)
43.9(a)	The person who does the work shall make an entry in the maintenance records containing: (1) a description of the work performed (or reference to a 337 or work order if applicable), (2) the date of completion of the work, (3) the name of person performing the work, and (4) if approved for re-

	turn to service, the signature and certificate number of the person who approved it. In addition, major repair and major alterations are to be entered on a separate form.
145.59(a)	A qualified inspector inspects before approval for return to service; the repair station certifies airworthiness and approves for return to service (Figure 5).
145.61	Repair station maintains adequate records of work it does, naming the person that does the work and the inspector of the work.

GATES LEARJET — TUCSON
FAA CERTIFIED REPAIR STATION NO. 462-41
TUCSON INTERNATIONAL AIRPORT, ARIZONA 85706

AIRCRAFT FLIGHT RELEASE

AIRCRAFT S/N _____
FLIGHT NO. _____
DATE _____

NO. OF OPEN FLIGHT SQUAWKS _____
NO. OF OPEN GROUND SQUAWKS _____
NO. OF OPEN REMOVAL ITEMS _____

I certify this aircraft has been inspected in accordance with the manufacturer's recommended _____ (*) inspection(s) and was determined to be in an airworthy condition.

(*) Enter: pre-flight, post-flight, daily, etc.

MECHANIC BADGE NO	INSPECTOR STAMP OR SIGNATURE	PILOT SIGNATURE (A/C ACCEPTED FOR FLIGHT TEST)

ADDITIONAL ITEMS OR REMARKS:

Figure 5. Inspection signoff–return to service (Courtesy of Gates Learjet Corporation)

Regulatory Requirements

43.11   Annual, 100-hour, and progressive inspection maintenance record entries.

The person approving or disapproving for return to service makes entry in the maintenance record including the following information: (1) type of inspection (and extent if a progressive), (2) the date and aircraft time in service, (3) the signature and certificate number of mechanic, (4) statement certifying airworthiness if approved, and (5) statement certifying unairworthiness if not approved.

Approved for Return to Service vs. Return to Service

43.7(a)   Persons authorized to approve for return to service as provided in FAR 65, 145, etc.

43.5(a)   No person may return to service an aircraft or article that has undergone maintenance unless: (1) it has been approved for return to service, (2) maintenance record entries have been made, (3) the major repair or alteration form has been executed, and (4) any change in operating limitations or flight data is revised and set forth as prescribed in 91.31.

91.31(a)   Operating limitations shall be complied with during operation.

91.31(b)   Operating limitations (aircraft flight manual, placards, listings, markings, etc.) must be current and available in the aircraft during operating, including: (1) power plant markings and placards, (2) airspeed markings and placards, and (3) aircraft weight and balance information and other data as required.

Approval for return to service is a maintenance record entry by an authorized individual.

Return to service is any action by any person to put an aircraft or article into an operational status after it has been maintained or altered.

FAR 121. Certification and Operations: Domestic, Flag, and Supplemental Air Carriers and Supplemental Air Carriers and Commercial Operators of Large Aircraft

121.369(b)(2)   This section and similar provisions of FAR 127 deal with the designation of maintenance and alteration work that must be inspected (required inspection items).

FAR 135. AIR TAXI OPERATORS AND COMMERCIAL OPERATORS OF SMALL AIRCRAFT

135.411(a)(2)   This section applies to maintenance requirements for aircraft type certificated for 10 to 30 passengers with a maximum allowable payload of 7,500 pounds.

135.411(b)   This section applies to maintenance requirements for a certificate holder operating aircraft for nine passengers or less under the operator's continuous airworthiness maintenance program.

FAR 147. AVIATION MAINTENANCE TECHNICIAN SCHOOLS

This part prescribes the requirements for issuing aviation maintenance technician school certificates and associated ratings.

The FAA is responsible for the regulation and promotion of civil aviation in such a manner as best to foster its development and safety. FAA aircraft maintenance responsibilities are accomplished by the general aviation and air carrier maintenance inspector in his or her day-to-day activities or certification and surveillance of aircraft, maintenance, airmen, and agencies. He or she is also charged with the investigation and reporting of aircraft accidents, incidents, malfunctions, defects, and violations of the FARs. Of the many various FAA job functions, the least desirable ones are accident investigation and violation investigation.

Responsible aircraft maintenance requires a cooperative effort on the part of all who have aircraft maintenance responsibilities.

The Boeing company is an outstanding example to consider when discussing the regulatory maintenance requirements of a carrier aircraft manufacturer.[4] The Air Transport Association of America listed in its pamphlet, *Air Transport, 1982,* 2,749 turbine aircraft.[5] In its membership airline fleet, 1,554 (727s [Figure 6], 737s, 747s [Figure 7]) of these aircraft were built by the Boeing Company. Of these aircraft, Boeing's 727 is the world's best-selling plane, with 1,832 orders since 1960.

Boeing's approach toward maintaining the airworthiness of an airplane starts during the design, development, and certification of a new airplane and continues throughout the aircraft's operational life.[6] Boeing states that "each airplane is a part of a system that requires a continuing dialogue between . . . the operator of the aircraft, the appropriate regulatory agency and the manufacturer".[7]

The airplane certification process conducted by the FAA is proof that the manufacturer has adhered to all existing regulations. The maintenance program developed at the time of certification is also an assurance that the

Regulatory Requirements 43

Figure 6. Boeing 727-100 (Courtesy of American Airlines)

airworthiness of the airplane is maintained while in service. After the airplane is delivered, the operator becomes the third party involved in maintaining continuing aircraft airworthiness. When a problem is discovered, a problem resolution and production change process is activated for both airplanes in production and in service. The maintenance program of the airplane will be revised as required. The FAA, the manufacturer, and the operator are all involved in the change process. Furthermore, the engineering and manufacturing operations have many controls to use in guarding against deviations from the type and production certification as shown in Figure 8.

Production Certificate—Reference: FAR Part 21, Subpart G. A production certificate must be applied for subsequent to the issuance of a type certificate and within six months of same. The applications forwarded by the prime manufacturer of the aircraft (Boeing) to the FAA Regional Office. Additionally, Boeing must submit information relative to its organizational structure, personnel, facilities, and equipment that will be utilized to manufacture the specified model airplane. A complete description of Boeing's quality control system is also required to be transmitted to the FAA. The system must include all aspects of the control of such items as inspection planning, production planning, materials review, inspection records, fabrication and assembly, technical data control, drawing change control, process control, tool and gauge control, purchasing and receiving, suppliers, storage and issuance, final assembly and test, and shipping and domestic delivery. The

Figure 7. Boeing 747 (larger aircraft) and Boeing 707 (smaller aircraft) on loading ramp, 1982 (Courtesy of American Airlines)

FAA then conducts sufficient audits and reviews to assure the validity of the submitted quality control system and upon acceptance of that system will issue a Production Certificate authorizing the manufacturing of production aircraft.

After the initial type certification of a given model, all follow-on production airplanes must continue to meet the requirements of the Type Certificate, or FAA approval of all significant changes from the previously certificated type design configuration is required. Figure 9 illustrates the sequence of events from change requirements to a Boeing airplane delivery.

Development of the initial maintenance program for a new airplane model involves all three of the principles previously noted. This activity,

Regulatory Requirements

while independent of the actual airplane certification, is carried on concurrently with its completion scheduled to support introduction of the new airplane into carrier service. As noted in Figure 10, specific requirements included in Federal Aviation Regulations (FAR) Part 25.1529 require the manufacturer to provide the purchaser of an aircraft with information considered essential for proper maintenance. FAR Part 121 contains the specific requirements with which a scheduled United States airline must comply in operating and maintaining the airplane. Advisory Circular 121-22 provides guidelines for establishing and conducting a Maintenance Review Board (MRB) on a newly manufactured aircraft, powerplant, or appliance to be used in air carrier service. In addition, FAA representatives participate in the manufacturer/operator meetings during the Maintenance Program Development. Maintenance experience of initial purchasers of the new airplane model, as well as the manufacturer's experience, are utilized in the development process.

The maintenance program development is managed by a Maintenance Steering Group (MSG) composed of airline and airframe and engine manufacturer representatives. The steering group is also supported by several specialist working groups. Each working group includes airline, manufacturer, and FAA specialists.

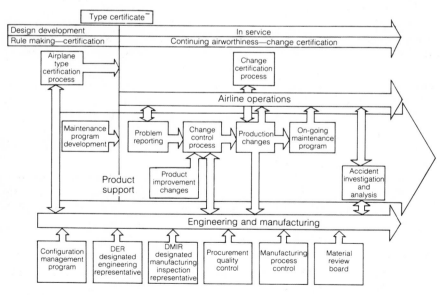

Figure 8. Maintaining continuing aircraft airworthiness (Courtesy of The Boeing Company)

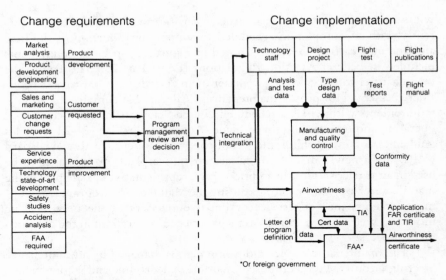

Figure 9. Production airplane certification process (Courtesy of The Boeing Company)

The manufacturer's role in development of a maintenance program for a new model airplane starts with the collection and evaluation of data from systems and structures design. This is followed by detail analysis using MSG decision logic procedures developed by the maintenance Steering Group. The MSG then establishes the required maintenance tasks and the type of maintenance process required. The individual requirements with comparable time intervals are packaged and developed into checks conforming to airline maintenance practices. Meetings between the manufacturer and customer airline finalize the maintenance program proposal. The FAA then appoints a Maintenance Review Board to evaluate the proposal, and approval is released in the form of an FAA MRB Report.

The Maintenance Review Board Report contains the minimum initial maintenance requirements required by the FAA that must be included in the airline's Operation Specification-Aircraft Maintenance. All MRB items, in addition to a great amount of additional information required to develop the airline's Detailed Maintenance Program, are included in the manufacturer's Maintenance Planning Document.

The airline's Maintenance Operations Specification is prepared in accordance with FAA Advisory circular 121-1A and is then submitted to the regulatory agency for approval. It is important to note that the maintenance program is approved by the regulatory agency specifically for the airline that

Regulatory Requirements 47

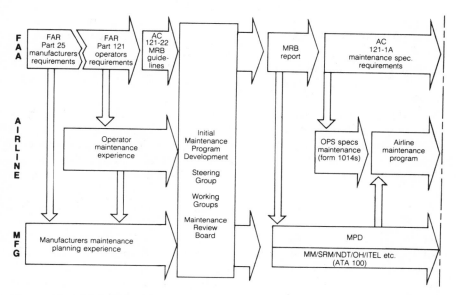

Figure 10. Initial maintenance program development (Courtesy of The Boeing Company)

has submitted it. It is based not only on the aircraft manufacturer's recommendations but also on the airline's experience, route structure, flight segment lengths, operational environment, fleet size, facility, and maintenance capability. Each operator has unique considerations requiring a specific aircraft customized maintenance program.

The Maintenance Review Board Report issued by the FAA and the Manufacturer's Maintenance Planning Data (Documents and Task Cards) are supplemented with a number of other publications and services that assist the airline in planning for maintenance and operation of the new aircraft. The manufacturer's other publications directly applicable to aircraft maintenance include the following: Maintenance Manual, Wiring Diagram, Overhaul Manual, Structural Repair Manual,[8] Corrosion Prevention Manual, Nondestructive Test Document, Illustrated Tool and Equipment Manual, and Airline Maintenance Inspection Intervals Report.

Once an airplane has been placed in operational service, maintaining its airworthiness is essential. A good problem-reporting system involving manufacturer, operator, and the FAA is required. The following section describes how Boeing, through its Field Service Engineering Organization, collects information on service problems, how the organization functions to ensure the resolution of these problems, and how the "loop is closed" with feedback to the operator and to the FAA. Federal Aviation Regulations (FAR) 21, Subpart A, Section 21.3, together with FAA Advisory Circular 21-9A, delineate the statutory requirements for the reporting of certain types of failures, malfunctions, and defects to the FAA.

The Customer Support communications network is utilized to report quality control problems that may be encountered on in-service airplanes. Special handling of all quality control items that occur during the first thirty days of service of each new airplane is performed by the Customer Support Representative. In these instances, the Customer Support Engineering Unit relays the report directly to the Quality Control Organization of the Product Division on a Quality Control Item Form. Receipt of this form requires a thorough and complete investigation of the relevant manufacturing and inspection processes and procedures to ensure elimination of the problem.

It can be seen that during the life of the airplane constant improvement is sought. Changes are made to eliminate problems that develop, improve safety, operational reliability, and maintainability. Airplanes are changed by adding new models, higher gross weights, different engines, and advanced technological equipment. Each change is examined to assure compliance with the rules and the FAA's specific requirements. This is a continuous process with each change properly scheduled so approval by the FAA may fit complex manufacturing and delivery schedules. This continuing process of change and improvement directly affects the ongoing airline maintenance plan. To the extent these changes are retrofitted to airplanes in service, the airline's maintenance programs will be affected as shown in Figure 11. The approval of changes made by the airline to in-service airplanes, as well as the related maintenance operations specification and supporting data, is the responsibility of the appropriate regulatory authority. The airplane manufacturer can recommend but has no authority to insist on or approve airline maintenance changes.

As with all aircraft manufacturers, those that design and build large aircraft, or those that concentrate on general aviation types, proof of airworthiness comes with the final inspection, and post/preflight shakedown. It must be remembered, however, that each manufacturer spends millions of dollars and thousands of hours to test intensively each new model well before final post/pre-flight procedures are carried out.

This chapter reviewed the basic regulatory documents imposed by the FAA on all aircraft manufacturers, owners, and operators. To show how compliance with the maintenance regulations is followed, the author utilized extensively the Boeing Company's excellent notes on their concept of "maintaining continuous aircraft airworthiness." Each major step in maintenance regulatory requirements is indicated, initiating with Boeing's airplane certification process and continuing through their service problem reporting

Regulatory Requirements

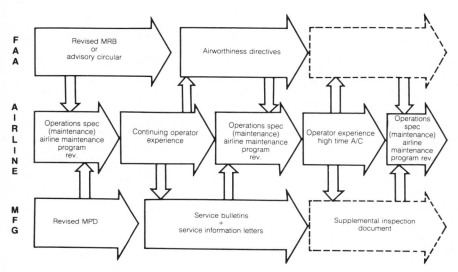

Figure 11. Ongoing maintenance program (Courtesy of The Boeing Company)

and their ongoing maintenance programs. It has been seen that the operator of the aircraft, the appropriate regulatory agency, and the manufacturer all contribute to maintaining airworthiness of the airplane in service. This is exactly the intent of the Federal Aviation Administration.

4

Organizational Structures

DUE TO THE COMPLEXITY, VARIETY, and extreme depths of the many different aviation organizations, it will be found that the formal organizational structure may take one or more of the accepted formats. Basically, there are three formats to consider: functional, line, and line and staff.

Functional organizations (overall structure) are the least used, although many small and new aviation activities tend to start this way. The functional approach simply means that all work areas are divided into specialty fields, i.e., auto/facilities maintenance, base maintenance, technical services, and others, with each field/area being supervised (Figure 12).

Line organizations normally indicate the managers, supervisors, and workers whose primary jobs are to see that things get done—in other words, the "doers" or the "action" operators. Typically, the workers are assigned to work in a designated unit, rather than a specific functional area, indicated in Figure 13.

The third type of organization consists of the line-and-staff concept. It was previously stated that those that did the work represented a line organizational concept. But often others are needed to help decide what to do, how to do it, provide services for line workers, and many other functions in the way of assistance. Departments that assist or advise line departments in controlling quality and maintaining records are typical staff activities. Exam-

Organizational Structures

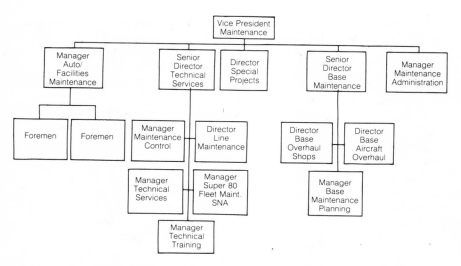

Figure 12. Example of functional organization: Frontier Airlines (office of vice president, maintenance), November 3, 1982 (Courtesy of Frontier Airlines, Inc.)

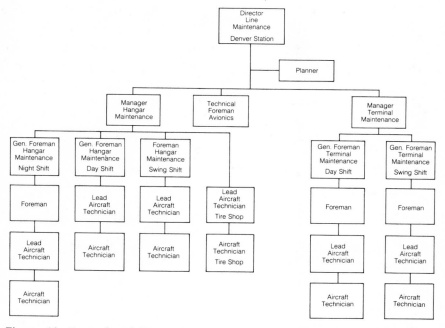

Figure 13. Example of line organization: Frontier Airlines (line maintenance, Denver station), November 3, 1982 (Courtesy of Frontier Airlines, Inc.)

ples of staff activities include industrial engineering, plant maintenance, finance, personnel, legal, and purchasing. Most line-and-staff organizations appear to be functional in concept and it is often hard to make the distinction. However, that aviation manager must be aware of the three major types of organization structures and what type will be most efficient for his or her particular activity.

Boeing's organization chart (Figure 14) shows a hybrid functional-line-staff[1] concept which is not unusual for this type of manufacturer. Their maintenance functions extend into every major division, but initial responsibility falls in the office concerned with airworthiness and safety. Service problems become the responsibility of the VP for customer services.

One major airline's maintenance and engineering organization structure is shown in Figure 15. The subsequent charts show combinations of function, line, and staff concepts. It is interesting to note the magnitude of the maintenance operation in this international carrier. Supplementing the mentioned airline structure is a generalized airline maintenance organizational chart shown in Figure 16.[2] Organizational data on one of the largest major airlines, United Airlines, Inc., is given in the following pages.

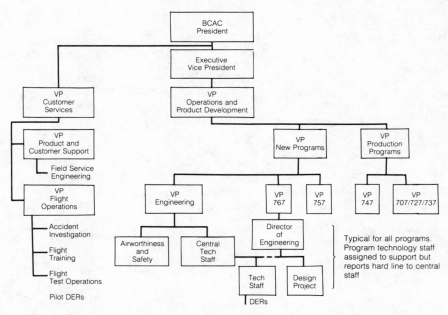

Figure 14. The Boeing Company organization chart (Courtesy of The Boeing Company)

Organizational Structures

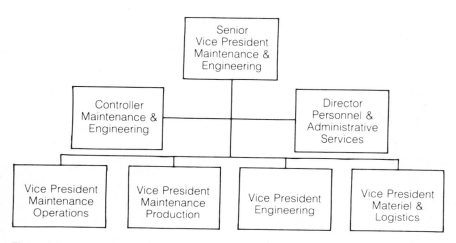

Figure 15. Maintenance and engineering organization of a typical major airline

As the nation's oldest air carrier, United Airlines has been identified with many firsts.[3] The company, for example, was first to fly fare-paying travelers coast-to-coast (1927), first to develop and adopt a practical system of two-way, plane-to-ground voice radio communications (1929), and first with all-cargo flights (1940).

United's extensive route system links all cities and 92 airports in the United States, Canada, and Mexico. The route system includes 9 of the top 10 traffic-generating centers and 20 of the leading convention cities in the United States. The company serves New England and the Atlantic Seaboard; Toronto, Ontario; Southeastern states and the Great Lakes region; Midwest and Mountain states; and Pacific Northwest; and Southwest states of Nevada, Arizona, Texas, and Oklahoma; the coastal area from Vancouver, British Columbia, to San Diego and the Hawaiian Islands. In April 1983, United began serving Tokyo and Hong Kong from the Pacific Northwest.

The 317-plane fleet includes 18 Boeing 747 and 47 McDonnell Douglas DC-10 wide-body aircraft for high density operations; 13 DC-8 Freighters and 29 Super DC-8s for long-range, nonstop operations, 154 Boeing 727s in standard and stretched configurations for short to medium routes, and 49 Boeing 737s for short-range routes. In September 1982, United introduced its first Boeing 767 new-generation twinjet on scheduled flights between Chicago and Denver, Boston and San Francisco. The aircraft was designed for fuel economy and operating efficiency: United has ordered 39 of these airplanes, with options on 30 more. The extent of United's operations requires unusual managerial techniques and vast communications systems,

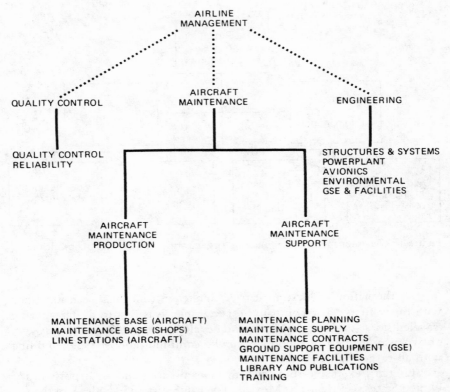

Figure 16. Generalized airline maintenance organization, June 1982 (Courtesy of Douglas Aircraft Company)

both telephone and teletype. The teletype circuits can handle 300,000 messages per day. This is supplemented by private-line telephone networks.

United traces its origins to Varney Air Lines, which began airmail service between Pasco, Washington, and Elko, Nevada, via Boise, Idaho, on April 6, 1926. Varney Air Lines later was merged into Boeing Air Transport, along with Pacific Air Transport and National Air Transport. Boeing Air Transport was part of a combine which included Boeing Airplane Company and the engine manufacturer, Pratt and Whitney. United Air Lines was organized in 1931 as the management company for the airline division. Three years later the combine underwent divestment and its corporate divisions became separate business entities.

The greatest increase in the company's size occurred on June 1, 1961, when Capital Airlines was merged into United. Capital had begun operations in 1927, carrying mail between Pittsburgh and Cleveland. The merger added some 7,000 employees and increased United's route system by 7,250 miles. In 1969, United, which today has some 41,000 employees, became the wholly owned subsidiary of a holding company, UAL, Inc. UAL, Inc., also owns Western Hotels and GAB Business Services, Inc.

United's executive headquarters, northwest of O'Hare Airport, Chicago, includes administrative offices, a training center, and reservations headquarters. The company's Flight Training Center is in Denver, its Maintenance Operations Center in San Francisco. Each day of the year in all seasons, the company operates more flights than there are minutes in the day. United carries over 50 million passengers yearly in addition to flying over 800 million cargo ton miles.

United Airlines' Maintenance Operations Center is the largest facility of its kind in the world. Located on a 140-acre site near San Francisco International Airport, the sprawling complex includes over three million square feet of floor space. It is the maintenance and modification center for the airline's large fleet of jetliners. Major responsibility of the MOC is the repair, overhaul, and modification of aircraft, engines, and several million component parts. It also provides the technical specifications, information, plans, and materials for all fleet maintenace processes. In 1982 more than 1,000 airframes were routed to the San Francisco facility for maintenance and special project work. In the same year, the MOC checked and repaired more than 1,400 engines, modules, and auxiliary power units. Over 6,000 employees—largest concentration at any of the airline's facilities—work at the Maintenance Operations Center, generating a payroll that totaled more than $264 million last year. Purchases of supplies and services amounted to $168 million in 1982.

Four major departments operate out of the sprawling complex (Figure 17). These are Aircraft and Engine Maintenance, Technical Services, Maintenance Administration, and Maintenance Supply. In addition, support organizations provide specialized services, such as medical, computer and communications, auditing, insurance, ground safety, investigation and security, credit union, and food services.

Work on aircraft, engines, and components is performed in accordance with company specifications, which are often more demanding than requirements of the Federal Aviation Administration. Disassembly, repair, and inspection of major aircraft structures, as well as engines and components, are governed by at least one of these company or government regulations. An elaborate inspection system is used to double-check all repair or modification work.

A wide variety of service shops take care of avionics equipment, flight instruments, electrical, hydraulic and pneumatic units; sheet metal, plastic, and wood components; and carpeting, draperies, seat covers, and other items

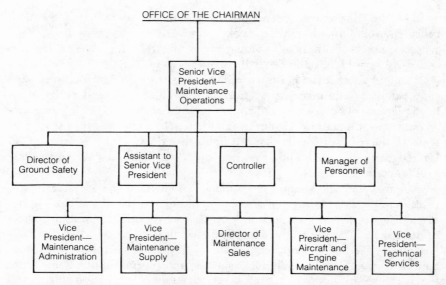

Figure 17. United Airlines maintenance operations, 1983 (Courtesy of United Airlines)

used in the aircraft cabin. Among the sophisticated repair techniques seen at the facility are electron beam welding, plasma spraying, high-pressure bonding, and nondestructive testing of aircraft components. Specialized equipment includes two modern Teleplatforms which feature a high degree of flexibility for aircraft access, five acoustically treated engine test cells, a massive autoclave for bonding large aircraft structures, test equipment incorporating the latest state-of-the-art technology, and a computerized materials distribution system.

Construction of the Maintenance Operations Center was started in 1940, when the airline's Western Operations staff moved from Oakland. During the postwar period, an almost entirely new facility was built to accommodate the new four-engine piston fleet. In the late 1950s, the plant was enlarged to provide for the jet fleet, and expanded again in the early 1970s to accommodate the wide-body jets. (See Figures 18 through 24 for detailed maintenance activity at the maintenance operations center.) Details of the United Airlines photographs are as follows:

Figure 18. United Airlines' maintenance facility at San Francisco International Airport is a technological center where the latest state-of-the-art equipment in aircraft maintenance is employed. The 140-acre complex is staffed by more than 6,000 technical specialists and support personnel who

Organizational Structures 57

Figure 18. United Airlines maintenance facility, 1983 (Courtesy of United Airlines)

maintain the airline's fleet of 300-some aircraft, including the new-generation Boeing 767 twinjets.

Figure 19. Inertial navigation equipment (shown in the lower left hand corner of the photograph) is tested by a mechanic at United Airlines' Maintenance Operations Center, San Francisco. the sophisticated electronic boxes, two (and sometimes three) of which are installed on each of the airline's 747 and long-range Douglas DC-8 aircraft, are routinely maintained by United's overhaul facility, where the units are analyzed on the console shown. Each year, an average of 400 inertial navigation boxes, each costing approximately $100,000, are serviced by skilled technicians in the airline's avionics shop.

Figure 20. A dramatic view from inside the exhaust funnel of the newest engine test cell at United Airlines' Maintenance Operations Center, San Francisco International Airport, shows mechanics working on a General Electric CF-6 turbine engine of the type that powers the Douglas DC-10 wide-body jetliner. The open half-shells are fan reverser cowls which remain closed in flight to provide a clean airflow and are opened to permit access to engine components during maintenance. United has five acoustically treated test cells at its maintenance facility where the airline, during 1982, repaired more than 1,400 engines, modules and auxiliary power units.

Figure 19. Inertial navigation equipment test, 1983 (Courtesy of United Airlines)

Figure 21. A Boeing 747 jetliner, sporting United Airlines' new stylized symbol and fuselage paint scheme, is towed out of a massive hangar following maintenance work at United's sprawling overhaul complex near San Francisco International Airport. The huge facility, spread over a 140-acre site, is the maintenance and modification base for the airline's 317-plane fleet. It is the largest repair base of its kind in the world.

Figure 22. Flanked by twin banks of high-intensity lights, a body panel from a Boeing 747 jetliner goes through a drying process after its aluminum skins have been bonded to the honeycomb core. The process is one of several sophisticated techniques employed at United Airlines' maintenance operations center, San Francisco International Airport, where the carrier overhauls some 1,000 airframes and 1,400 turbine engines each year.

Figure 23. Looking like a dart that hit its target, the drive shaft in the combustor section of a General Electric CF-6 engine is one of the components that will receive thorough inspection by a mechanic at United Airlines'

Organizational Structures

Figure 20. Inside the exhaust funnel: engine test cell, 1983 (Courtesy of United Airlines)

maintenance operations center, San Francisco International Airport. Some 1,400 turbine engines, including the CF-6, which powers the McDonnell Douglas DC-10 wide-body jetliner, are overhauled at the airline's 140-acre maintenance facility each year. Subject of the mechanic's scrutiny are the engine's fuel nozzles.

Figure 24. A time exposure presents molten slag as filaments of light during an inert gas welding operation at United Airlines' maintenance operations center, San Francisco International Airport. The mechanic is working

Figure 21. Maintenance hangar for United Airlines' wide-bodied aircraft, 1983 (Courtesy of United Airlines)

on an internal oil sleeve from a turbine engine which powers the airline's wide-body aircraft.

Facts and figures on United's Maintenance Operations Center. Mission: Provide safe and reliable airplanes, operating procedures, and support equipment for United's operations at the lowest total cost relative to the competition and consistent with the new corporate mission.

1982 statistics (Maintenance Operations Division): Over 6,000 employees at the Maintenance Operations Center, and at 13 line stations around the system; $264 million payroll, including benefit costs; 140 acres with over three million square feet of floor space; 12 aircraft docking positions; 5 modern engine test cells; and the biggest, most sophisticated airline maintenance facility in the world.

Maintenance of aircraft and equipment at the line stations is accomplished by maintenance services personnel in each geographic division. Maintenance Operations specifies what, how, and how often maintenance must be accomplished, and provides technical support 24 hours a day, 7 days a week for line stations through direct contact with maintenance controllers in San Francisco.

Organizational Structures 61

Figure 22. High-intensity lights: drying process, 1983 (Courtesy of United Airlines)

Four major organizations of United's Maintenance Operations Division:
1. Aircraft and Engine Maintenance—accomplishes major maintenance, repair, and inspection of airframes, engines, components, and ground power units. (3,400 employees; more than 1,000 airframes, 1,400 engines, modules, and auxiliary power units, and over one million aircraft components maintained in 1982; approximately 14,000 manhours spent on each major widebody aircraft overhaul.)
2. Technical Services—provides technical expertise and quality audit of all aircraft and ground equipment maintenance. Monitors quality of maintenance tasks performed. Works with aircraft, engine, and component manufacturers in development of new equipment, and therefore responsible for technically supporting this equipment in service (800 employees).
3. Maintenance Administration—provides maintenance planning, analysis, information, and plant services for the Maintenance Operations Division. Provides support services for the line stations through Line Maintenance Control staff (515 employees).
4. Maintenance Supply—responsible for supporting the aircraft fleet

Figure 23. General Electric CF-6 engine inspection, 1983 (Courtesy of United Airlines)

Organizational Structures

Figure 24. Inert gas welding operation, 1983 (Courtesy of United Airlines)

and the Maintenance Operations Division with spare parts, equipment, supplies, etc. Maintains storeroom facilities at the Maintenance Operations Center and line stations (650 employees; controls $350 million inventory, including spare engines; provides 90% availability of parts at storeroom windows; and purchases approximately $150 million worth of goods and services per year).

A commuter airline, illustrated by the organizational chart shown in Figure 25, represents a commuter airline with an above-average annual growth rate for the last several years. The airline currently has 13 aircraft

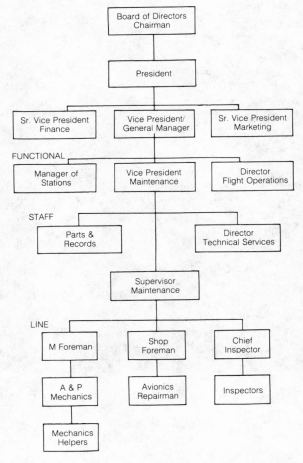

Figure 25. Organizational structure of a typical commuter airline

Organizational Structures

(9 Embraers and 4 Beechcrafts) and is home-based at a Louisiana city regional airport. Maintenance is accomplished by the airline at only two stations. (The airline flies into 14 stations.) They are certificated as repair stations in avionics only. Due to FAA regulations, the airline must have a fully documented maintenance program and all mechanics must be A&P licensed. Since the airline's aircraft are designed to carry 10 or more passengers, the FAA requires continuous inspection procedures. The total organizational structure, as presented in Figure 25, shows the maintenance activity reporting to the vice-president for maintenance. Even though this airline is considered small, the airline's structure represents a combination of function, line, and staff.

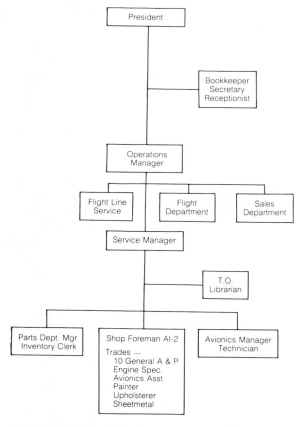

Figure 26. Typical fixed base operator organizational chart

Depicted as an example of a general aviation fixed base operator is an actual independent fixed base operator (FBO) located at an Arizona city airport. The purpose of the company is the selling of aircraft and aviation services. In this endeavor it is similar to an automobile dealership. However, the nature of the aviation activity has added more critical areas to the scope of the operation. Aircraft charters, flight instructions, and FAA regulations put much more demands upon the maintenance department. The organizational chart (Figure 26) shows a basic line-and-staff operation with approximately 50 percent of the total company employees working under the service manager, where the primary maintenance functions are performed.

It has been the intent in this chapter to show the organizational structures of four varied types and different sized aviation activities: a major aircraft manufacturer, a major airline, a small but progressive commuter airline, and a privately owned fixed base operator. The difference in the size, scope, and aviation activity tells the story in the organizational structures. But it is also interesting to note that in all four companies, there were several similarities in their organizational concepts.

5
Management Responsibilities

THIS CHAPTER INTRODUCES THE STUDENT to the complexities of aviation maintenance management. It pertains to the successful managing of maintenance operations by being involved in the five essential management functions of planning, organizing, staffing, directing, and controlling.[1] These management functions are applicable to any type of facility or operation being maintained; the methods presented can be used to improve the efficiency and to reduce the costs of maintenance operations for activities involved in air transportation, airline operations, aircraft manufacturing, or general aviation management.

During the last decade, a number of technological changes have evolved that have assisted in bringing the decision-making process into sharper focus.[2] From the age of manually operated decision processing to the current extensive use of high-speed computers and data processing procedures, the quantity and timeliness of information has added immeasurably to that available to aviation managers. They can now base their decisions on far greater amounts of relevant data than previously.[3]

Planning is the function of establishing and clarifying goals and objectives. Planning involves looking ahead and making decisions as to what one is going to do and preparing a plan, the blueprint for future action. Of all the management functions, planning is the one that permits a maintenance department to act rather than react. Thus, establishing objectives is the first requirement (in other words, management's anticipated accomplishments). Then the company policy is established because it is the practice or method used to achieve the objectives.

Planning then develops into the identifications of short-range and long-range objectives and the determination of how to achieve them. This is, as stated above, through a predetermined course of action, frequently called a master plan.[4] An action plan must be prepared detailing the course of action and stating the specific resources required. This becomes the company's strategy. Strategic decisions[5] are normally major decisions regarding the utilization of resources, both material and human, that may have a long-term impact on the accomplishment of goals or objectives. Prior to looking toward a short-range or long-term requirements is the maintenance manager's responsibility to determine first the current situation. He or she must evaluate the operation's assets and liabilities. He or she must look into population trends, transportation costs, sales trends or competing companies, share of the market nationally, statewide, and locally, the potential for continuing leasing or buying, and governmental regulations that could put limitations on the business's growth.

Developing first the long term goal involves consideration of the time span in years. Appropriate long-range goals should be for no less than 3 to 5 years with succeeding 5 year intervals but not beyond a 10 to 15 year total time span. For these time periods, specific goals (always written) should include total income, size of operation, capital investment (renovation, replacement, new purchase) on buildings and equipment, replacement and retirement of permanent maintenance employees, and geographical sales scope.

Short-range plans are easier to develop than long-range plans due to fewer uncertainties and a shorter time span. The short-range plan requires, however, some specific plans that would contribute toward to the achievement of the long-range plans. There are a number of realistic goals that can be considered during this short time span: adding service capability, building more maintenance hangers, increasing spare parts items, expanding sales activity, cutting costs, increasing maintenance and administrative efficiency, expanding flight school attendance, expanding transient activity, and many others. It must be remembered that, for each specific goal to be attained, a larger segment of the maintenance budget may be required.

Further, if plans are not expressed in terms of time, they are useless. Target dates are essential to the accomplishment of work load. The maintenance mechanic should always know how much time is allocated to complete a given task. Calendar time is the measure essential to proper planning.

Management Responsibilities

Planning as a managerial activity normally includes the use of budgeting, which is expressed in terms of time as well as money.

One of the prime dangers is for the manager, in developing and completing his or her plan, to expect the plan to remain static with no changes required. It is almost certain that the plan will require changes from time to time, in many cases, due to changing economic conditions, personnel demands, cost increases in spares, and other unforeseen happenings—in other words, to be realistic, plans must be flexible.

Scheduling is one segment of planning that specifies given objectives to be accomplished in relation to time. Schedules can and should be applied to as many maintenance operations as possible. For example, inherent in any preventive maintenance program is the use of schedules to delineate when and how long it will take specific types of maintenance to be performed. The use of schedules as a planning device is not restricted to the maintenance department. Schedules may also be used as a coordination and communication means in dealing with other departments. For example, as in the case of upcoming purchase orders, service contracts renewal, and maintenance vehicle registration dates renewal. Effective planning requires quantification—unless plans are defined in terms of numbers as well as time and money, they will probably remain useless.

The management function of organizing is the development of the organizational structure and authority relationships necessary to achieve determined objectives. Saying it another way, organizing establishes the positions within an organization and assigns tasks to each. For a maintenance department, organizing is the grouping of activities necessary to achieve the objectives of the department and the assignment of a supervisor for each group. In arriving at a structure suitable for the work activity involved, the manager should be responsive to the basic types of formal organization structures and the advantages of each. The types to consider are functional, line, or line-and-staff. These were reviewed in some detail in Chapter 4, "Organizational Structures."

Proper performance of organizing can lead to substantial cost savings in material, labor, and capital investment in maintenance shop tools and equipment. Additionally, each maintenance department should customize its organization to most efficiently and profitably meet its own particular mission. There is no one optimum way to organize, but an important starting point from the standpoint of cost and performance is, first, determine what work has to be performed, and second, decide what work belongs together.

When a maintenance activity increases in size, a structured organization must be created, consisting of several groups of coordinated activities, each managed by a supervisor. At this point, the maintenance department has to consider three basic groupings or a combination of these if the work activity and plant size warrant it. These groupings are: (1) by operation,[6] (2) area or zone maintenance shops, and (3) central shop.

Organization by operation would consist of separate groups for such activities as inspection, preventive maintenance, repair, and major overhaul. Airline maintenance normally follows the organization by operation approach since airlines' concern is with specific types of equipment, i.e. airplanes, engines, etc.[7]

Area maintenance shops (or zone maintenance shops) are organized in such a way that assignments of each maintenance group or individual are to regularly assigned areas or locations. This type of organization is considered decentralized. There are several reasons for having regularly assigned maintenance area mechanics: (1) The travel time to and from the central shop is greatly reduced. (2) There is sufficient work to keep the group or individual busy every day. (3) The group or individual usually takes a greater interest in the specific assigned area. (4) The requirement for special skills to maintain specific equipment takes considerable time to learn.

A central shop is a location in a plant or hanger where a craft group is headquartered. A shop may consist of one or more crafts responsible to the shop supervisor. The rationale behind the central shop concept lies in ease of administration, availability of better equipment, improved training and on-the-job (OJT) considerations, improved central control, and ease of task assignments. The layout of the maintenance activities should be analyzed continually to determine how much centralization or area (zone) maintenance is required.

Even though the maintenance activity is considered a line organizational concept, the use of staff personnel as support is increasingly followed by the larger aviation companies. Support or staff activities may be established in a stockroom operation, purchasing service, tool crib, finance function, personnel services, transport services, work order processing, and planning staff.[8] The primary reason for staff functions in support of the line maintenance workers is to allow the line personnel to concentrate on their specialized work with minimum interruptions.

Organizing establishes the organization's positions and details the tasks to each. In staffing, the manager has the responsibility for recruiting, hiring, training, and developing the right person for each job. These responsibilities also include defining, initially, the manpower requirements. This is a never-ending managerial responsibility because all companies lose personnel through retirements, promotions, transfers, layoffs, resignations, and deaths.

There is a valid need for a well-defined staffing function in the larger maintenance department. Without one, it would be difficult to carry out the

Management Responsibilities

continued profit-making activity of the company. Also, personnel efficiency will wane unless staffing controls are placed into the system to maintain the required competence. Effective staffing requires that the maintenance manager observe all personnel performance, with the intent of increasing work efficiency through the avenues available to him or her, i.e. training, education, OJT, transfers, dismissals, and the like.

There has been and will continue to be a requirement for the maintenance department to hire and keep qualified mechanics and helpers. Assuring appropriate compensation and the maintenace of a continuous training program are requisites of the staffing function for all maintenance personnel, experienced or new. The determination of manpower requirements involves a knowledge of existing work load, future work load (already on the books or forecast), availability of skills in the marketplace, and current required skills. The skilled maintenance manager will draw on his or her managerial experience and historical shop records to determine needs for the present and the future. In addition, the manager can utilize computer expertise to assist in making the right manpower decisions. The size of the current maintenance department will have a lot to do with what approach the manager uses to justify and fulfill his or her needs. Several methods[9] are examined by competent technological managers and consultants in various treatises. Foremost among these approaches or methods are the informal, planned, and formal levels. Within these methods and based on work force size lies the use of the historical basis, the adjusted historical basis, level-of-effort planning, and factor analysis.

The use of job descriptions and aptitude tests is a widespread approach toward selecting potential mechanics. Larger aviation firms, as well as all branches of the federal government, make extensive use of the applicant's résumé, scores on aptitude tests, and personal interviews to determine the selection of interested candidates.

Basically, the need for filling any skill-required maintenance position depends upon the level of maintenance required by the activity. Normally, the requirements center around knowledge related to the airframe, power plant, electronics, and propeller. FARs 65 and 147 include specific skill requirements for these and other aviation tasks. Finally, in making the selection(s), look for such things as: "PERSERVERANCE, as demonstrated by a work record that shows the applicant can stay with a difficult situation; ALERTNESS, as indicated by the applicant's ability to follow your description of the work to be done . . . ; COOPERATION, as illustrated by the applicant's willingness to go through the red tape of the employment interviewing and processing without quibbling about it."[10]

Organizing calls for assigning duties to individuals; directing is the implementation of this structure. In the organizational process of directing, the maintenance manager is called upon to exercise skills in the areas of leadership, motivation, communication, counseling, and discipline.[11] The directing function involves the ability to guide and motivate maintenance personnel so as to achieve the goals of the organization, and at the same time build a work environment that will keep the workers happy and content in their jobs and in their relationship with the company. Confidence and understanding by maintenance department personnel for their manager is a necessary ingredient for successful management, but this can only be accomplished by management's careful direction in imposing the tasks to be done.

Aside from the formal authority given the maintenance manager, he or she must also exert leadership to motivate subordinates to contribute willingly their expertise in accomplishing high-quality work. Leadership is defined as "the quality of a leader: capacity to lead."[12] What this definition means is that the leader or manager must have the ability to persuade or influence others in doing what is necessary to meet the objectives of the maintenance department and the company.

Not everyone is the same type of leader. He or she may subscribe to one of these three basic leadership styles: the "autocratic" leader who makes decisions without consulting anyone; the "democratic" leader who invites subordinates to participate in the decision-making process; and the "laissez-faire" leader who lets the subordinates make many of the decisions. Since managers are dealing with human beings, no one of these styles is always the most effective but the successful manager has, more often than not, selected correctly the best approach for each individual decision.

One essential of good leadership is the ability to motivate people, and the best way to introduce the subject of motivation and incentives is to apply real-world conditions in relating the background of an aviation activity and its problems. In some respects, the whole operation at an aviation company is similar to a training ground for maintenance personnel.

Changing jobs is common in the aviation industry because of the broad pay scales involved and the experience levels required. The pay and experience levels are normally directly proportional to the cost and complexity of the aircraft. The best that a small company (FBOs and other repair stations) can hope for is to hire an outstanding individual right out of A&P school, try to keep him or her for approximately two years; then give the person a good recommendation as he or she leaves for bigger and better things. Once in a great while, an exceptional individual may be found who is capable of handling a supervisory position and who is willing to remain to gain management experience.

Webster's Dictionary says to motivate means "to provide with a motive." A motive is an incentive that causes a person to act. No one to date, however, has the absolute answer to the problem of work motivation. The behavioral scientists, who deal with human action and seek generalizatons of man's behavior in society, are looking into the theory of basic human needs for possible answers to the work motivation problem. The prominent behavioral scientist Abraham H. Maslow has suggested that there is a hierarchy of needs that extends in this order; physiological (lowest—includes the basic needs to survive), safety, social, esteem, and self-fulfillment (highest need). Maslow says that after each ascending need fulfillment, the next highest need becomes dominant. Management would do well to have a better understanding of the theories projected by the scientists who have spent their careers making a study of human behavior and work motivation. (Bell Helicopter does an excellent job in bettering and understanding human behavior through its management seminars.)[13] In addition, a manager must have an instinct for dealing with subordinates. Some call this an inherent liking for human beings. Certainly what is required is an undestanding of the people with whom the manager must deal.

Communication is another essential facet of good management and an element of the directing function. A number of studies have been made substantiating that the greatest amount of time spent in communicating is spent in face-to-face, oral communications.[14] The overall activity of communicating includes the means by which all maintenance personnel exchange work information, future requirements, costs and improvements in operational processes, and numerous other concerns affecting companies' productivity. Effective communication may be written or oral; if oral, the communication intent may have to be followed up in writing. Maintenance managers have a significant need for clear, well understood communications. Yet it is usually in the areas of communications that minor work problems develop between working personnel and supervisors. Communication will fail when managers use authority rather than understanding in obtaining compliance with their directions.

In the complexities of any organization, there are many barriers which cause the message to lose its true meaning or become lost in the transmission. To overcome some of these barriers that inhibit or block effective communications, a communication skill called corrective feedback may be used. The corrective process of feedback is a continuing effort to find out if the message sent or received is fully understood. When corrective feedback is operative, the received message is fully understood. The receiver of the message is consciously trying to determine if what he or she received is what the sender wished to send. The sender is trying to find out from the receiver whether what was sent has been received as he or she wanted it received. Just as a servo mechanism is used in a guided missile system to keep a missile on course, corrective feedback is used to keep the message on track.

There are two types of feedback—one negative and the other positive. A negative feedback occurs as follows: A worker asks his or her supervisor a question and the supervisor answers, "That's a good question," but never gives an answer. Another type of negative feedback occurs when an order is given and the individual charged with the task asks for instruction but the sender walks off because he or she is too busy. Positive feedback occurs when a person is given a job to do and the feedback between the sender and the receiver is so clear that the receiver walks away feeling confident enough to do the job that the sender has assigned.

Feedback is one of the most valuable tools that a manager can use in communication. Feedback should never be used to threaten, condemn, or critize. Feedback of this nature can result in blocking of communication, or developing an increasing hostility on the part of both parties. Feedback should be used in asking people to tell you specifically what they think you have said. Quite often a supervisor gives out orders and occasionally they are not always accomplished the way the supervisor had intended. Asking others to tell you what they think you have said leaves less room for error.

In communications there is also external feedback and internal feedback. Internal feedback is when we ask ourselves whether this is what we meant to say and external feedback is when we are concerned with what others have said.

The *Harvard Business Review* on management suggests that company managers tape-record their decision-making meeting.[15] This is especially necessary for those meetings that consist of difficult decision-making. The tape would then be held until a later time for the decision-making group to listen again to the tape of the previous meeting. Listening to the tape would become a learning experience. These tapes could also be very useful to another person or consulting firm from whom an unbiased opinion and fresh new ideas could be solicited.

To sum up, maintenance managers correctly utilizing the function of directing and applying good communication techniques will instill in their subordinates a sense of confidence and help to create work enthusiasm. Good communication techniques will assist in developing worker loyalty and integrity and in the delegation of work responsibilities as appropriate. Finally, these techniques will assist in the spirit of well-being in the maintenance activity.

The intent of management's controlling function is to assure that current progress and future goals are still "on-time" in the planning category. If there is evidence of schedule slippage, the controlling function must investigate and take the necessary action to again reach the programmed schedule intervals. Plans are useless if no corrective actions are instituted to correlate and implement the on-time requirements established by written objectives and time schedules.

Management Responsibilities 75

Since management control consists of assuring that current progress is conforming to plans, there has to be a close working relationship between personnel charged with the planning function and those people responsible for controlling. The essential elements of an effective management control system consist of: (1) developing detailed plans and procedures, (2) collecting and reporting data on work performance through some specified period, (3) analyzing and comparing actual versus desired progress, and if necessary (4) taking corrective action to bring the program back on schedule.

In actuality, controlling starts with the planning activity. Thus, in the development of any good plan, yardsticks for measuring progress and comparisons will be used. These yardsticks will be in terms of time and money and frequently both. Techniques used to assist in the control measurements (time and/or money) are: labor hours, labor costs, labor standards,[16] budgets schedules, CPM (Critical Path Method), PERT (Program Evaluation Review Technique), automation by operations systems, and others.

As with the directing function, the first step in the control cycle is the result of planning activities. Here, too, lies close coordination effort between control and planning if the entire plan development is to succeed. A major source of concern will be in the process of adjusting future progress based on an analysis of past performance data. The controlling efforts will quickly determine the validity and accuracy of the initial plans and will raise the question as to whether new plans should be formulated or the original plans modified.

When initiating the discussion on incentives, it is best to remember that not all workers will accomplish a full day's work for a full day's pay. This really provides the basis for incentives. Maintenance workers are not unlike workers in other industries. A statistical number of them will perform below a reasonably attainable level. Much of this can be attributed to management's failure to provide conditions under which normal or improved performance levels can be reached. There is also another responsibility that management has—providing motivation, as discussed earlier in this chapter.

Some maintenance managers concern themselves with assuring minimum machine down-time, and with good work relationships between maintenance and other departments. This, of course, is fine from the standpoint of developing the confidence and self-esteem of the maintenance department. It is also good for the motivation of the workers. What should be wanted, however, is some incentive arrangement that will motivate the workers to push a little harder or to go that "one step further."

Whatever the reason for having an incentive plan, it must be well designed and well administered to be successful. Changes to the plan, to the detriment of the worker, will create worker-motivation obstacles and probably ensure the plan's failure. Incentive plans that fail do so due to: insufficient coverage, changes that affect the monetary gain, poor or inadequate administration, loosely designed plan, lack of top management support, and poor communication.

A good strong incentive plan pays attention to the needs of the employees as well as the company. The people want an accurate evaluation of their performance (normally the weakest link in the plan), incentives that are attractive but reasonable, an increased earning opportunity for everyone, and a plan that the workers can understand. The company wants higher production with continued high quality, safety, control, and good industrial relations.[17] Since World War II, there has been a far-reaching technological breakthrough in automation, resulting in an increased need for maintenance skills. Together with automation increases in the production industries, the aviation industries have grown tremendously, furthering the need for even higher-grade maintenance skills.

While an incentive plan is thought of as a means for compensating employees related to output or quality or both, the incentives themselves may take the form of financial or nonfinancial arrangements. Regardless of what management texts say about financial reimbursements in the form of salary increases or otherwise, the financial inventive is a powerful one. Nonfinancial incentives appeal to the workers' pride, emotion, or ego but do little to sustain the superior work or continued production that is wanted by management.

Financial incentives take the range of pay increases, merit ratings, profit sharing, bonuses, compensation time awards, additional time off, and other features. Such plans may be for individual efforts or for a group working together. The method decided upon will depend on the work environment or the product or process involved. The basic purpose, however, remains the same—to promote a strong motivation force, accomplished through extra compensation, to achieve higher company objectives, and in turn, higher company profit. The bottom line is achievement as high as realistically possible.

To give an idea of what a good incentive plan will do is to compare a normal day's work with one where an incentive exists. Based on a 100 percent as a fair day's work level (standard), industry has been measured at 70 to 85 percent. With incentives in being, the work force will average out at 110 to 125 percent. After incentive payments, the work force still will show an increase at a level of 100 to 110 percent—a substantial improvement over the norm. Additional bonuses for management usually include less spoilage, higher quality, and even better attitudes on the part of the employees. Incentives can also be a major factor in making maximum use of a limited work force, particularly where higher skills are required as in the typical aviation maintenance facility.

Citing a major airline as an example of motivation through an incentive contract, the airline operates on a "commitment" basis. All supervisors and managers have an input to decide how many hours are required for their shop to do a certain job. They are then held to that commitment. Commitments are revised each year to what the supervisors and managers feel they can live with. Each worker takes pride in his or her work and endeavors to produce in order that the commitment can be realized. The employees do this committed work under a labor contract. The incentive program has been a success in that the company has been making a substantial profit for several years and employee turnover is only 1.5 percent per year, an important factor in aviation maintenance operations.

Successful applications of incentives requires many things but foremost is adequate preparation, communications, and proper timing. Initially, top management should hold meetings to discuss what they have in mind regarding an incentive plan. Supervisors and managers should have an input before the final plan is "set in concrete." Proper timing and good communications down to and including the workers is paramount. For example, a typical incentive program is going to reduce the work force by some percent, if effective. Therefore, timing on implementing the plan is critical. Optimum timing would be while the department is receiving more work orders than they can handle with the existing personnel. Finally, good planning and scheduling are part of the necessary prerequisites to a successful maintenance incentive program.

The functions of planning, organizing, staffing, directing, and controlling are keys to the success of the manager's activity. To be a success in the management field, supervisors, like their subordinates, require intensive training and development programs. The skills of the manager in administration, human relations, and technology have to be honed to a fine point. The manager's effectiveness rests on these three skills. The manager must be alert to the theories of behavioral scientists such as Maslow and attentive to Maslow's hierarchy of needs. These managerial concerns, if developed and adhered to, followed by the practice of trying to understand subordinates and their unique work motivation problems will make a more efficient and highly desirable company manager.

6

Aviation Maintenance Procedures

THIS INTRODUCTION WILL CALL ATTENTION to those maintenance requirements specified by appropriate FARs and the inspection procedures called for. (FAA requirements and forms are listed in Appendix B.) For example, Appendix A of Part 43 details what constitutes major alterations, major repairs, and preventive maintenance. Regarding inspections, Appendix D of Part 43 indicates the scope and detail of items (as applicable to the particular aircraft) to be included in annual and 100-hour inspections. Since the glossary (Appendix C) gives many definitions and the chapter on government regulations also contains some maintenance concepts, the discussion here will continue without a review of such terms as preventive maintenance, major repairs, and so forth.

Initially, the discussions in this chapter will involve a detailed review of a continuous airworthiness maintenance program, aviation maintenance procedures that are requirements for essentially all types of aircraft.[1] This type of review will cover a considerable portion of the aviation maintenance activity. Additionally, an outline of an airline's maintenance manual will follow, showing the correlation between regulations (FAA) and applications (airline).

Managers operating maintenance departments subject to the provisions of (FAR) Parts 121, 127, or 135[2] are required to maintain continuous airworthiness maintenance programs. Basically, the FAR provisions point to the privilege of performing maintenance, inspections, and alterations. Thus, the ensuing discussion will deal with the elements of a continuous airworthiness maintenance program and the management areas associated with those programs, namely: responsibility for airworthiness, maintenance and inspection organization, performance and approval of maintenance and alterations, arrangements for maintenance and alterations performed by other persons, and continuing analysis and surveillance.

Before getting into the program elements, one term, "airworthy," needs to be clarified. The term is not defined in the Federal Aviation Act or the regulations; however, a clear understanding of its meaning is essential for use in the FAA's enforcement program.[3]

A review of case law relating to airworthiness reveals two conditions must be met for an aircraft to be considered "airworthy." These conditions are:

1. The aircraft must conform to its type design (certificate). Conformity to type design is considered attained when the required and proper components are installed and they are consistent with the drawings, specifications, and other data that are part of the type certificate. Conformity would include applicable supplemental type certificates and field-approved alterations.

2. The aircraft must be in condition for safe operation. This refers to the condition of the aircraft with relation to wear deterioration. Such conditions could be skin corrosion, window delamination/crazing, fluid leaks, tire wear, etc.

A continuous airworthiness maintenance program is a compilation of the individual maintenance and inspection functions utilized by an operator to fulfill total maintenance needs. Authorization to use continuous airworthiness maintenance programs is documented by Operations Specifications—Aircraft Maintenance, approved by the Federal Aviation Administration, for each user as provided for by FAR 121.25, 121.45, 127.13, and 135.11. These specifications prescribe the scope of the program, including limitations, and they reference manuals and other technical data as supplements to these specifications. Following are the basic elements of continuous airworthiness maintenance programs:

Aircraft Inspection. This element deals with the routine inspections, servicing, and tests performed on the aircraft at prescribed intervals. It includes detailed instructions and standards (or references thereto) by work forms, job cards, etc., which also serve to control the activity, and to record and account for the tasks that comprise this element.

Scheduled Maintenance. This element concerns maintenance tasks performed at prescribed intervals (see Figure 27). Some are accomplished concurrently with inspection tasks that are part of the inspection element and may be included on the same form. Other tasks are accomplished independently. The scheduled tasks include replacement of life-limited items, components requiring replacement for periodic overhaul, special inspections such as X-rays, checks, or tests for on-condition items, lubrications, etc.

Unscheduled Maintenance. This element provides procedures, instructions, and standards for the accomplishment of maintenance tasks generated by the inspection and scheduled maintenance elements, pilot reports, failure analyses, or other indications of a need for maintenance. Procedures for reporting, recording, and processing inspection findings, operational malfunctions, or abnormal operations such as hard landings, are an essential part of this element. A continuous aircraft logbook can serve this purpose for

MODEL			SERIAL NO.		REGISTRATION NO.
INTO WORK DATE			A/C DUE OUT DATE		
DESCRIPTION			AIRFRAME TOTAL TIME		
INTERVAL		APPROVAL	DESCRIPTION		
150	300	MECH	INSP		
				A. ELECTRICAL	
X	X			1. Landing and taxi lights.	
X	X			2. Anti-collision beacons and navigation lights.	
X	X			3. Pitot tube and angle-of-attack vane heaters.	
X	X			4. Static port heaters.	
X	X			5. Angle of attack vanes for shaker-pusher and indicator operation. (Refer to Chapter 27.)	
X	X			6. Cabin and cockpit lights.	
X	X			7. Instrument, radio panel, and glareshield lights.	
X	X			8. Visual and aural warning systems.	
X	X			9. Ice detection lights for operation. (Refer to Chapter 30.)	
X	X			10. Nose wheel steering. (Refer to Chapter 30.)	
X	X			11. Flaps, spoilers, horizontal stabilizer, and trim tabs for proper operation and accurate indication.	

Figure 27. First page of scheduled maintenance form (Courtesy of Gates Learjet Corporation)

occurrences and resultant corrective action between scheduled inspections. Inspection discrepancy forms are usually used for processing unscheduled maintenance tasks in conjunction with scheduled inspections.

Engine, Propeller, and Appliance Repair and Overhaul. This element concerns shop operations which, although they encompass scheduled and unscheduled tasks, are remote from maintenance performed to the aircraft as a unit. As with the aircraft scheduled and unscheduled elements, instructions and standards should be provided along with means for certifying and recording the work. Appropriate life-limited parts replacement requirements are included in this element.

Structural Inspection Program/Airframe Overhaul. This element concerns the structural inspections identified as the D and E check level by Maintenance Review Board reports and/or airframe major overhaul. As with the aircraft inspection program, detailed instructions and standards should be provided along with a work control and recording means. In addition to structural inspection, airframe major overhaul programs schedule extensive maintenance tasks.

Required Inspection Items (RII). This element concerns maintenance work items which, if improperly done or if improper parts are used, could endanger the safe operation of the aircraft. RII items appear in all elements of the operator's continuous airworthiness maintenance program. They receive the same consideration regardless of whether or not they are related to scheduled or to unscheduled tasks; i.e., the fact that an RII requirement arises at an awkward time or at an inconvenient location has no bearing on the need to accomplish it properly. (1) There are many tasks throughout each continuous airworthiness maintenance program which, although not in the RII category, are essential to a safe, reliable, and efficient aircraft. A responsible maintenance program specifies inspection of these tasks to ensure their proper accomplishment. The operator should designate the tasks that need to be inspected as a general requirement to assure the effectiveness of his or her program as well as the RII items. (2) The operator should identify required inspection items on work forms in a suitable manner. For example, such items may be identified with the abbreviation "RII," an asterisk, or any workable method.

The operator's maintenance manual[4] serves to define the continuous airworthiness maintenance program and to provide procedures and instructions for its use. It is comprised of three general categories: (1) policies and procedures, (2) detailed instructions for the accomplishment of the scheduled inspection program, and (3) technical manuals for maintenance standards and methods.

1. The policies and procedures segment deals with organizational matters, the policies of the maintenance section, procedures for the administra-

tion of the continuous airworthiness maintenance program, test flight requirements, and many other subjects that are peculiar to each individual operator. It is a company publication and serves as an administrative tool for directing and controlling the total maintenance function and to define all facets of the maintenance operation and their interrelationship. Quality control is a major subject of this publication.

2. The segment of the maintenance manual system dealing with the scheduled inspection program is usually a company publication. It normally includes the work forms or job cards associated with scheduled inspections and detailed instructions (or specific references) for accomplishing the inspections.

3. Technical manuals concern how to accomplish specific tasks. They set forth methods, technical standards, measurements, operational tests, etc. These are usually manufacturers' publications, the applicability of which is designated by the policy and procedures manual. Technical manuals can be supplemented by the operator.

FAR Section 121.363 and corresponding sections of FAR Parts 127 and 135 afford the following maintenance privileges to operators subject to these regulations: (1) to perform maintenance, preventive maintenance, inspection, repairs and alterations on the aircraft they operate; and (2) to develop (or adopt) a continuous airworthiness maintenance program and to tailor and adjust that program and related practices and procedures to best suit the operator's need.

FAR 121.365, 127.132, and 135.423 impose organizational requirements with regard to the administration of the continuous airworthiness maintenance program. This does not mitigate the applicability of FAR 43 nor does it waive initial aircraft certification requirements. The Required Inspection Item (RII) requirement causes the operator to separate the inspection organization from the remainder of his or her maintenance organization to ensure proper accomplishment of RII items. This separation applies to the following functions: RII items performed by the operator's organization, means to ensure RII items performed by other persons are subjected to RII inspection separation by the other person's organization and procedures, and identification of RII items by a means that is understood by the person performing the work.

The operator is privileged to perform maintenance on his or her aircraft in accordance with his or her continuous airworthiness maintenance program and for other operators under corresponding parts of the Federal Aviation Regulations in accordance with their programs.

The operator's manual prescribes the authorizations, methods, standards, and procedures for performance of that maintenance. This is recognized by FAR 43.13(c).

The operator's aircraft are released for service (airworthiness release, ref.: FAR 121.709, 127.319 or 135.443) following maintenance by a person specifically authorized by the operator rather by an individual or repair station on their own behalf.

When an operator uses the services of another person to accomplish all or part of his or her continuous airworthiness maintenance program that person's organization becomes, in effect, an extension of the operator's organization. The operator should execute contractual agreements with the persons performing his or her work on a continuing basis to ensure the operator's interests are met. In the case of major operations such as engine overhaul, the agreement should denote a specification for the work and that specification should be included or referenced as part of the operator's manual system.

FAR Part 121.373 and similar provisions of FAR Parts 127 and 135 require the operator to provide a system for continuing analysis and surveillance of his or her continuous airworthiness maintenance program including work performed according to that program by another person. This requirement, in effect, establishes a quality control or internal audit function.

This system should provide for timely corrective action on the following: (1) frequency of unscheduled parts replacement or need for unscheduled maintenance, (2) degree and frequency of adjustment and calibration of equipment, and (3) changes in operational capability or reliability (delays, etc.).

This system should provide a continuous audit of the total maintenance system to assure that everyone connected with it is in compliance with the operator's manuals and the applicable regulations. This should include, but not be limited to, the following: (1) all publications and work forms are current and readily available to the user; (2) maintenance is, in fact, performed in accordance with the methods, standards, and techniques specified in the operator's manuals; (3) maintenance forms are screened for completeness and proper entries, and RII identification; (4) records pertaining to tracked components are cross-referenced to stock-issue records, etc., to minimize errors; (5) indications of inadequate training; (6) airworthiness releases are executed by designated persons and in accordance with the procedures specified in the operator's manuals; and (7) carryover items and deferred maintenance are properly handled.

A maintenance facility inspection[5] may be defined as any inspection made for the specific purpose of determining the adequacy of personnel and facilities at any base, terminal, or intermediate stop along the route flown by an air carrier at which maintenance is to be performed on aircraft operated by the air carrier. In the certification of an air carrier, inspection of maintenance facilities should be accomplished prior to the time proving flights are conducted and prior to the start of operation in the case of supplemental air carriers and commercial operators.

In conducting a maintenance facility inspection, the primary objective is to determine that adequate housing, equipment, spare parts, technical information, and qualified personnel are available to perform satisfactorily the functions that are to be accomplished at that particular location. In the case where the inspection of required inspections are to be performed, the operator must have a separate maintenance and inspection organization. He or she must have properly trained, qualified, and authorized personnel to perform such inspections, and he or she must maintain a list of these individuals.

Operations Specifications, FAA Form 1014 (OMB 04-R0075), are issued to supplement air carrier and air taxi rules by listing authorizations and limitations that are not specifically prescribed by the regulations.[6] FAR Sections 121.25(b)(6), 121.45(b)(6), 127,13(b)(7), and 135.11(b)(2) specify that time limitations for overhaul, inspections, and checks be set forth in the operations specifications. Operations specifications may also authorize optional privileges afforded for by FAR Section 21.197. Some aircraft have parts that are life limited by the manufacturer that must be listed in or referenced by the aircraft operations specifications. When approved, the provisions of the operations specifications are as legally binding as the regulations themselves. (Ref.: FAR Sections 121.3, 127.11 and 135.5).

Operations specifications fall into eight broad categories, each of which is referred to as a "part." Each part has an assigned letter designator as follows:

1. Part A—General
2. Part B—Enroute authorizations and limitations
3. Part C—Airport authorizations and limitations
4. Part D—Maintenance
5. Part E—Weight and balance
6. Part F—Scheduled cargo flights, charter flights or other special services
7. Part G—Equipment interchange
8. Part H—Aircraft leasing

The progressive inspection system has been designed to provide scheduling of inspections of aircraft on a predetermined basis. This system is particularly adaptable to larger multiengine aircraft and aircraft operated by companies and corporations where high utilization is demanded.[7] The inspector should not attempt to establish arbitrary intervals for inspection or overhaul of aircraft. Intervals should be based on the manufacturer's recommendations, field service experience, malfunction and defect history, and the type of operation in which the aircraft is engaged.

An air taxi maintenance base inspection is any inspection made for the specific purpose of determining the adequacy of maintenance personnel and maintenance facilities at any base or terminal where maintenance is performed on aircraft operated by an air taxi operator. Inspection of the operator's aircraft and hazardous materials program is included in the base inspection. This inspection determines if the operator is safely handling and transporting dangerous articles and magnetized materials (hazardous materials) in accordance with pertinent regulations.

Hazardous Materials Inspections. Hazardous materials inspections are performed in accordance with practices and procedures contained in Order 8000.34A, Transportation of Hazardous Materials. (1) Inspectors should be thoroughly familiar with FAR Part 135; the Code of Federal Regulations (CFR), Title 49, Subpart C, Parts 171 through 178; and current advisory circulars, legal interpretations, notices, alert bulletins, and Order 8000.34A. (2) Inspectors should determine that ATCO[8] personnel are familiar with 49 CFR Part 175 and that personnel are, or will be trained in the safe handling and carriage of dangerous articles and magnetized materials.

Maintenance spot and aircraft ramp inspections[9] are means of sampling the quality of maintenance and the degree of compliance with established maintenance standards and procedures.

Spot and ramp inspections should cover a variety of areas and range in scope from an inspection of an entire aircraft to a review of a particular maintenance function. If a weakness is known or suspected in a particular phase of maintenance, emphasis should be placed in these areas.

Spot and ramp inspections should be conducted with a minimum of interruption to maintenance operations and passenger handling. The inspector will advise the person in charge of the affected maintenance as to his or her (the inspector's) mission and area of interest. He or she will normally discuss activities with mechanics and inspectors but will address complaints or the need for corrective action with appropriate supervisory or management personnel.

Maintenance spot inspections are the primary surveillance activity used by maintenance and avionics inspectors for certificate holders under FAR Parts 121, 127, and 135. They are observations and analyses of in-progress maintenance operations for overall quality, conformity to the operator's inspection or maintenance programs, compliance with specified methods, techniques, and practices, competency of personnel, and adequacy of facilities. All of this is viewed from the standpoint of the effectiveness of the operator's management control and support.

Spot inspections should consider the following: (1) availability of, and compliance with the policies, procedures, and practices published in the operator's manuals or other technical material applicable to the work in progress; (2) adequacy of facilities support equipment relative to proper accomplishment of the work; (3) competency of personnel, including indication of adequate training; (4) parts availability; (5) execution of paperwork and that proper work forms are used and accounted for; (6) work environment including disruptions due to job reassignments, shift change continuity; (7) currency of test equipment calibration dates; and (8) compliance with RII items (as appropriate).

Spot inspections of scheduled aircraft checks or inspections should be programmed and should consider, in addition, the: (1) quality of inspection writeups and related corrective action entries; (2) use of procedures to negate or modify inspectors' writeups with particular regard to concurrence by the quality control organization; (3) control and accountability of routine and inspection writeup forms; (4) availability of, and use of, instructions and standards for work being performed; (5) control of deferred and carryover items; and (6) production and quality control turnover procedures.

Spot inspections should also be programmed for shop activities.

Spot inspections should be accomplished on work performed by contract agencies. A particular consideration is the relationship between the operator and the agency regarding the responsibilities of each party and accountability for those responsibilities.

Ramp inspections are inspections of in-service aircraft in the operational environment. Their purpose is to determine the maintenance of the aircraft by direct inspection rather than by evaluation of in-progress maintenance. Observation of refueling, passenger handling, and ground equipment usage is normally accomplished during the check.

Guidelines to follow during maintenance ramp inspections:

1. Maintenance Manual.
    a. Copy aboard if required. Note revision concurrently.
2. Aircraft logbook.
    a. Pilot complaints.
    b. Correction of service difficulties.
    c. Chronic mechanical difficulties.
    d. Carryover items.
    e. Inspection time limits.
3. Exterior of Aircraft.
    a. Fuselage.
    b. Wings.
    c. Control surfaces.
    d. Empennage.
    e. Wheels and tires.
    f. Landing gear.
    g. Leaks—fuel, oil, hydraulic.
4. Interior of Aircraft.
    a. Seats.
    b. Seatbelts.

c. Placards.
d. Signs.
e. Emergency equipment.

The above-ramp outlines are cues to general areas of coverage. The individual inspector will use the cue as a tool for prompting his or her memory in light of general knowledge and knowledge of a particular airplane and/or operator. Examples might be as follows:

The Cue	*What the Cue Might Bring to Mind*
Seats	1. That in all aircraft the security of seat attachment to the structure is of paramount importance.
	2. In airplanes using floor rails, both the rail-locking device and the seat-attachment fitting are subject to breakage. The seat floor rails often become filled with debris which prevents proper locking.
	3. High-density loading often involves breakaway back seats which are subject to maintenance and location problems.
Oxygen	1. Is the capacity and pressure up to standards?
	2. Are masks clean, protected, and sufficient in number?
Emergency Exits	1. Is there a sufficient number of exits for the number of seats installed? Are any exits blocked by seats or equipment?
	2. Are they placarded properly and legibly?
	3. Do the emergency lights illuminate the exits sufficiently and properly?

The preceding examples are only a few of the infinite number of significant items that can be cued by the ramp inspection outlines. Each make and model of airplane has its own problem areas and, upon inspection, these areas should be checked first. Time and circumstance will control the extent of the inspection toward the ultimate or complete inspection. In any case, the problem areas or suspect areas should always be checked.

Aircraft Type Certificated for Nine or Less Passengers.[10] This section concerns management personnel requirements, aircraft inspection programs, and additional maintenance requirements for air carrier and operating certificate holders for aircraft they operate that are type certificated for nine or less passengers with a maximum allowable payload of 7,500 pounds under the provisions of FAR Section 135.411(a)(1).

The aircraft to which this section applies are subject to: (1) Part 91 (general operating and flight rules) inspection programs, (2) an approved aircraft inspection program (AAIP) when requested by the operator or required by the FAA, and (3) a continuous airworthiness maintenance program if the operators select this option under FAR Section 135.411(b). The inspector should evaluate the certificate holder's operation and aircraft type involved to determine if the program submitted by the operator is appropriate and adequate.

The additional maintenance reqiurements rule applies to the engine, propellor, rotor, and emergency equipment. It does not apply to the airframe and is not intended to impose a continuous airworthiness maintenance program nor to augment otherwise the aircraft inspection program.

FAR Section 135.421(a) provides for two types of additional maintenance programs for engines, propellers, rotors, and emergency equipment for the aircraft type model involved: (1) Authorization to use an aircraft manufacturer's maintenance program and/or the individual product manufacturer's program. This arrangement is normally selected by new operators and operators who are incapable of, or not otherwise interested in, developing their own additional maintenance programs. (2) Individually approved programs developed by the operator for his or her use. This provision is for operators capable of developing and analyzing a maintenance program for effectiveness and revising it accordingly.

Paragraph (b) of Section 135.421 defines manufacturer's maintenance programs. The term "maintenance instructions" includes service bulletins, letters, or other publications concerning maintenance applicable to specified models and configurations (modification status or other groupings that influence maintenance needs). Publications dealing with repairs, alternations, or other matters beyond the scope of the term "maintenance" are not required by this FAR section but may be included in the additional maintenance program to support higher maintenance intervals or other inspection variables. The term "manufacturer's instructions" does not include individual authorizations or recommendations by a repair facility or manufacturer to a particular owner or operator.

Adoption of a manufacturer's maintenance program in toto by an operator is authorized by the rule. Approval for its use by an operator concerns only the following: (1) its applicability as to make, model, configuration, etc.; and (2) that the program is sufficiently comprehensive to satisfy the rule. If the aircraft manufacturer's program does not include engine overhaul (or comparable heavy maintenance) and the engine manufacturer's program does, the operator should be required to designate the engine manufacturer's program to the degree necessary to encompass the engine overhaul requirement. It may be necessary to designate service bulletins or other manufacturer's maintenance instructions, in addition to a maintenance manual, in order to delineate an adequate program. If there is any doubt as to what limitations are imposed, the operations specifications or programs referenced thereon should explicity denote these limitations.

Operator-developed programs differ from adopted manufacturers' programs in the following respects: (1) An operator-developed program bears no prior FAA approval. It is incumbent on the operator to justify the program. If the program is, in effect, a manufacturer's maintenance program with operator variations, then it is an operator-developed program, not an adoption of a manufacturer's program. (2) All changes to a program of this type require FAA approval. Changes to a manufacturer's program should not be automatically incorporated into an operated developed program but they should consider it for incorporation.

An Airline Maintenance/Inspection Procedure. A typical airline procedure for accomplishing its maintenance and inspection requirements usually follows a chronological order in line with appropriate FARs. The maintenance manual reflects the organizational structure and the work to be done. A maintenance manual outline would cover the following basic aspects:

Maintenance organization
1. Structures
2. Policy and aims

Personnel policy and procedures
1. Introduction
    a. General rules and regulations

Operations specifications (Part D)

Maintenance and engineering responsibility—general
1. Accident and reportable incident procedures
2. General
    a. Practice

Manual requirements—general
1. Introduction

Aircraft reliability assurance—general practices
1. Airworthiness directives
    a. Standards
    b. Policy
    c. Practices

Maintenance and preventive maintenance—general
1. Scheduled maintenance program

Maintenance and preventive maintenance—inspection
1. Reguired inspections
    a. Standards
    b. Policy
    c. Practice

Maintenance and preventive maintenance—aircraft maintenance
1. Maintenance practices—operator
   a. Policy
   b. Practices
2. Scheduled maintenace
   a. Policy
   b. Practices
      (1) Production planning
      (2) Flight control
      (3) Maintenance control
      (4) Stores
      (5) Aircraft maintenance
3. Incompleted work procedure
   a. Policy
   b. Practices
      (1) Maintenance foremen
      (2) Lead mechanic
         (a) Class I and Class II maintenance stations (See "Maintenance Station Classifications and Procedures" subsequently, this chapter.)
         (b) Class III maintenance stations
4. Inoperative equipment identification and recording
   a. Standard
   b. Policy
   c. Practices
      (1) Placard installation
      (2) Placard removal

Maintenance and preventive maintenance—maintenance recording
1. Completed records
   a. Standards
   b. Policy
   c. Practice
2. Return of aircraft to service following maintenance
   a. Standards
   b. Policy
   c. Practices
      (1) Unscheduled maintenance
      (2) Scheduled maintenance

In accordance with FAR 121.105, an air carrier is required to show that competent personnel and adequate facilities and equipment (including spare parts, supplies, and materials) are available at such points as are necessary for the proper servicing, maintenance, and preventive maintenance of airplanes and auxiliary equipment.

For example, the classification of Frontier Airlines maintenance stations is determined by the availability of necessary parts, equipment, and qualified

Aviation Maintenance Procedures

personnel to perform the type of maintenance as specified. When necessary, Service, "A", and "B" Maintenance Checks may be accomplished at any station on Frontier Airlines System and at off-system airports where the aircraft has landed due to weather conditions or emergency situations. If maintenance personnel and parts needed for the particular check due are not available at the station involved, they will be sent from the Denver Maintenance Base or some other qualified station where parts and personnel can be spared. Stations will be classified as to a particular type of aircraft.

MAINTENANCE STATION CLASSIFICATIONS AND PROCEDURES
(COURTESY OF FRONTIER AIRLINES, INC.)

Classes

	I	II	III
B-737	Denver	Dallas-Ft. Worth Kansas City Tucson	Albuquerque Salt Lake City Lexington, Ky. Santa Ana, Calif.
DC-9	Denver		

Reports and records—reports
1. General information
2. Recording information
   a. General
      (1) Standards
      (2) Policy
      (3) Practices
3. Scheduling maintenance
   a. Work scheduling
      (1) Standards
      (2) Policy
      (3) Practices
         (a) General
            (i) A/C visits
            (ii) Components
         (b) Procedures—A/C visits
4. Aircraft maintenance
   a. General
      (1) Standards
      (2) Policy
         (a) Flight and maintenance log
         (b) Nonroutine maintenance work card
         (c) SCEPTRE deferred work record
         (d) Discrepancy sheet

Maintenance and engineering department methods—material and work handling
1. Parts kit procedures
   a. Aircraft check kits

Maintenance and engineering department methods—CAB accounting
1. CAB account numbering
   a. General
   b. Account numbering
      (1) Station
      (2) Function
      (3) Objective
      (4) Equipment
      (5) Budget number assignments

Maintenance and engineering department methods—requisitioning
1. Capitalization
   a. Aircraft
   b. Buildings
   c. Maintenance equipment and tools
   d. Office equipment

This chapter reviewed the FAA requirements for maintenance and inspection for all types of aviation maintenance activities. A clarification of the term "airworthy" was introduced due to certain legal concerns when maintenance is accomplished. Types of inspections and inspection procedures were also discussed; and, finally, a detailed outline of what an airline's maintenance manual might contain was presented.

7

Applications of Aviation Maintenance Concepts

THE FIRST CHAPTER IN THIS VOLUME covered an introduction to aviation, followed by chapters on the Fedral Aviation Administration, FAA regulatory requirements, typical aviation industry organizational structures, management responsibilities, and maintenance procedures.

This chapter will be devoted to the application of the previously reviewed materials and FAA regulations. First to be considered, in the area of maintenance, will be the several FAA requirements that the manager is confronted with: (1) FAA compliance and related guidance documents, (2) maintenance certification, (3) personnel qualifications, and (4) inspections.

There are literally dozens of FAA and related documents that provide control and guidance over new and existing aviaiton facilities. In the area of maintenance facilities operation, however, this list can be narrowed down to a few such as:

  Part 21.   Certification Procedures for Products and Parts
  Part 39.   Airworthiness Directives (ADs)
  Part 43.   Maintenance, Preventive Maintenance, Rebuilding and Alteration
  Part 65.   Certification: Airmen Other Than Flight Crewmembers
  Part 145. Repair Stations
  Part 147. Aviation Maintenance Technician Schools

Appropriate maintenance advisory circulars
General aviation airworthiness alerts
Manufacturers' service handbooks and instructions
Technical standard orders (TSOs)
Type certificate data sheets
Parts manufacturing approval
Supplemental type certificates (STCs)
FAA field approval

Chapter 3 outlined major maintenance areas that pertained to the regulations listed above. However, more specifically, consider Part 43, Paragraph 43.3(a): "Except as provided in this section, no person may maintain, rebuild, alter, or perform preventive maintenance on an aircraft, airframe, aircraft engine, propeller, or appliance to which this part applies. Those items, the performance of which is a major alteration, a major repair, or preventive maintenance, are listed in Appendix A."

In Appendix A of Part 43 major alterations pertain to wings, tail surfaces, fuselage, engine mounts, landing gear, and so on. Thus, it is seen that an aviation maintenance facility must be aware of all pertinent FAA regulations and must be certified in order to be able to perform.

In order for any aviation facility to maintain a qualified maintenance program, it must conform to FAA's Part 145—Repair Stations. This part prescribes the certification requirements for obtaining a repair station certificate and associated ratings to facilities (Paragraph 145.1). It also prescribes the general operating rules for the holders of those certificates and ratings. Part 145 further indicates that a manufacturer of aircraft, aircraft engines, propellers, appliances, or parts thereof, may be issued a repair station certificate with a limited rating under Subpart D of this part.

For a maintenance facility to qualify for a repair station certificate, the operator must adhere to the following requirements (Paragraph 145.11): First, he or she must submit a statement of his or her reasons for wanting a repair station at his or her place of business. Then the operator must furnish two copies of a suitably bound brochure, including: physical description of facilities (with photographs), description of the inspection system, an organizational chart, names and titles of managing and supervisory personnel, and a list of services obtained under contract, if any. Then, the operator must submit a copy of his or her inspection procedures manual, a list of the maintenance functions to be performed for the facility, under contract by another agency, and, in the case of an applicant for a propeller rating (Class 1, 2, or 3), a list, by type or make, as applicable, of the propeller or accessory for which the operator seeks approval.

A domestic repair station certificate or rating is effective until it is surrendered, suspended, or revoked (Paragraph 145.17). A foreign repair

Applications of Maintenance Concepts

station[1] certificate or rating expires at the end of 24 months after the date on which it was issued. The holder of a repair station certificate must display the certificate and ratings at a place in the repair station that is normally accessible to the public and is not obscured (Paragraph 145.19). The certificated repair station must also allow FAA authorized personnel to inspect it, at any time, to determine its compliance with Part 145.

Domestic repair station ratings that may be issued under Part 145 are (Subpart B, Paragraph 145.31):

1. Airframe
   a. Class 1: Composite construction of small aircraft
   b. Class 2: Composite construction of large aircraft
   c. Class 3: All-metal construction of small aircraft
   d. Class 4: All-metal construction of large aircraft
2. Powerplant
   a. Class 1: Reciprocating engines of 400 horsepower or less
   b. Class 2: Reciprocating engines of more than 400 horsepower
   c. Class 3: Turbine engines
3. Propeller
   a. Class 1: All fixed pitch and ground adjustable propellers of wood, metal, or composite construction
   b. Class 2: All other propellers, by make
4. Radio
   a. Class 1: Communication equipment
   b. Class 2: Navigational equipment
   c. Class 3: Radar equipment
5. Instrument
   a. Class 1: Mechanical
   b. Class 2: Electrical
   c. Class 3: Gyroscopic
   d. Class 4: Electronic
6. Accessory
   a. Class 1: Mechanical
   b. Class 2: Electrical
   c. Class 3: Electronic

In addition, the FAA may issue a limited rating to a domestic repair station that maintains or alters only a particular type of aircraft component (i.e., airframe, powerplant, propeller, etc.) or performs only specialized maintenance requiring equipment and skills not ordinarily found in regular repair stations (Paragraph 145.33).

An applicant for a domestic repair station certificate and rating must comply with housing and facility requirements, as well. These requirements are (Paragraph 145.35): housing for the facility's necessary equipment and material; facilities for properly storing, segregating, and protecting materials, parts, supplies; space for the work for which the applicant seeks a rating; suitable assembly space in an enclosed structure where the largest amount of assembly work is done; provide suitable ventilation for the applicant's shop, assembly, and storage area; and provide adequate lighting for all work being done.

To fulfill special housing and facility requirements (Paragraph 145.37):

An airframe rating applicant must provide suitable permanent housing for at least one of the heaviest aircraft within the weight class of the rating sought.

Either a powerplant or accessory rating applicant must provide suitable trays, racks, or stands for segregating complete engine or accessory assemblies from each other during assembly and disassembly.

A propeller rating applicant must provide suitable stands, racks, or other fixtures for the proper storage of propellers after being worked on.

A radio rating applicant must provide suitable storage facilities to assure the protection of parts and units that might deteriorate from dampness or moisture.

An instrument rating applicant must provide a reasonably dust-free shop if the shop allocated to final assembly is not air conditioned.

The maintenance manager, responsible under Part 145, must provide adequate personnel who can perform, supervise, and inspect the work for which the station is to be rated (Paragraph 145.39). The repair station is primarily responsible for the satisfactory work of its employees:

The number of maintenance employees may vary according to the type and volume of the repair station's work.

The repair station manager shall determine the abilities of his or her supervisors and shall provide enough of them for all phases of maintenance activities.

Apprentices or students used in critical aircraft operations must be directly supervised in groups of not more than 10 unless integrated into groups of experienced workers.

The repair station manager must be appropriately certificated as a mechanic or repairman under Part 65 and must have had at least 18 months of practical experience in the procedures, practices, inspection methods, materials, tools, machine tools, and equipment generally used in the work for which the station is rated.

At least one manager in charge of maintenance functions for a station with an airframe rating must have had experience in the methods and procedures prescribed by the FAA for returning aircraft (including engines) to service after 100-hour, annual, and progressive inspections. This manager can be recommended by the station to the FAA and can be certificated as a repairman.[2]

Applications of Maintenance Concepts

A repairman certificate may be issued to maintenance personnel employed by certified repair stations, commercial operators, or air carriers.

The inspection authorization is the next level of responsibility for the aviation mechanic. Inspectors require additional experience and related factors to qualify for this position: the inspector must have been a certified mechanic for at least three years (the last two years active in maintenance of aircraft); have a fixed operations base; have the tools, equipment, shop space, and maintenance data available in order to inspect properly aircraft; and pass a written test on the ability to inspect according to appropriate FAA regulations.[3]

Since the inspection authorization expires on 31 March of each year, a renewal is necessary. The inspector must show at each renewal date (Paragraph 65.93) that he or she: performed at least one annual inspection for each 90 days of his or her authorization, performed two major alteration or major repair inspections for each 90 days of his or her authorization, and performed or supervised and approved at least one progressive inspection.[4]

This section of the chapter will involve a review of certain types of inspections, together with the inherent maintenance responsibilities.[5] First, under Paragraph 91.163, the owner or operator of an aircraft is primarily responsible for maintaining that aircraft in an airworthy condition, including compliance with Part 39.

Maintenance bulletins[6] provide inspectors in all specialty areas with information concerning conditions found on specific aircraft or components that may assist inspectors in the performance of their duties. Maintenance bulletins are issued for: (a) alerting all inspectors to situations or conditions that are potential safety hazards; (b) informing inspectors of unusual or unique inspection procedure; and (c) advising inspectors of findings which, while not considered to be potentially hazardous, are of such a nature that knowledge of them will enable inspectors to perform more competently their surveillance activities.

Standard Numbering System. ATA Specification No. 100 establishes a standard for the presentation of technical data by an aircraft, aircraft accessory, or component manufacturer. This specification uses a conventional numbering system with a two-digit chapter number applied for each major aircraft system as follows:

21. Air Conditioning
22. Auto Flight
23. Communications
24. Electrical Power
25. Equipment/Furnishings
26. Fire Protection

27. Flight Controls
28. Fuel
29. Hydraulic
30. Ice and Rain Protection
31. Instruments
32. Landing Gear
33. Lights
34. Navigation
35. Oxygen
36. Pneumatic
37. Vacuum
38. Water/Waste
49. Airborne Auxiliary Power
51. Structures
52. Doors
53. Fuselage
54. Nacelles/Pylons
55. Stabilizers
56. Windows
57. Wings
61. Propellers
65. Rotors
71. Powerplant
72. (T) Engine Turbine/Turboprop
72. (R) Engine Reciprocating
73. Engine Fuel and Control
74. Ignition
75. Air
76. Engine Controls
77. Engine Indicating
78. Engine Exhaust
79. Engine Oil
80. Starting
81. Turbines
82. Water Injection
83. Accessory Gearboxes

Numbering of Bulletins. Each bulletin pertaining to a particular aircraft system is numbered as follows: ATA Code, dash, and bulletin number. For example, Bulletin 21-1 is the first bulletin on Air Conditioning (Code 21); 57-4 is the fourth bulletin on Wings (Code 57).

With regard to the maintenance required, each owner or operator of an

Applications of Maintenance Concepts

aircraft shall have that aircraft inspected as prescribed in Subpart D (Part 91) or Paragraph 91.169, as appropriate, and shall, between required inspections, have defects repaired as prescribed in Part 43. In addition, he or she shall ensure that maintenance personnel make appropriate entries in the aircraft and maintenance records indicating the aircraft has been released to service (Paragraph 91.165).

Regarding inspectons, no person may operate an aircraft unless, within the preceding 12 calendar months, it has had (Paragraph 91.160): an annual inspection in accordance with Part 43 and has been approved for return to service by a person authorized by Paragraph 43.7, and an inspection for the issue and of an airworthiness certificate. No person may operate an aircraft carrying any person for hire, unless within the preceding 100 hours of time in service it has received an annual or 100-hour inspection[7] and been approved for return to service in accordance with Part 43, or received an inspection for the issuance of an airworthiness certificate in accordance with Part 21. The 100-hour limitation may be exceeded by not more than 10 hours if necessary to reach a place at which the inspection can be done.

Concerning altimeter system tests and inspections, no person may operate an airplane in controlled airspace under IFR, unless, within the preceding 24 calendar months, each static pressure system and each altimeter instrument has been tested and inspected and found to comply with Appendix E of Part 43.

On progressive inspections, the frequency and detail of the progressive inspection shall provide for the complete inspection of the aircraft within each 12 calendar months and be consistent with the manufacturer's recommendations, field service experience, and the kind of operation in which the aircraft is engaged. The progressive inspection schedule must ensure that the aircraft at all times will be airworthy and will conform to all applicable FAA aircraft specifications, type certificate data sheets, airworthiness directives, and other approved data (Paragraph 91.171).

The above described inspections pertain to requirements that apply to all United States–registered civil aircraft operating within or without the United States and where maintenance, preventive maintenance, and alterations are performed (Paragraph 91.161). Table 7 shows a review of certain required inspections with appropriate part subparagraphs noted.

Based on the FAA's requirements for all facets of maintenance and inspection, each air carrier, commuter airline, air taxi, air travel club, air cargo line, corporate aircraft, training facility and school, industrial application, and privately owned aircraft develops, or has developed for it, its own individual specified and required program. Let's take a look at several ongoing and different programs developed by the following aviation activities.

TABLE 7
TYPES OF INSPECTIONS

Subparagraph	Inspection Type	Inspection Program Required	Inspection Authorized (IA)
91.169	100-hour/annual	Yes	Yes
91.171	Progressive	Yes	Yes
91.217	Applies to twin turbine large multiengine	Yes	Must have certificated A&Ps
135.411(a)(1)	9 passengers & less	Yes	Yes
135.411(a)(2)	Continuous maintenance 10 passengers & above	Maintenance program	Must have certificated A&Ps

Source: Randy Brooks, General Aviation District Office (GADO), Shreveport, LA., 1983.

Frontier Airlines: Frontier Airlines,[8] whose corporate headquarters is based in Denver, Colorado, also has its only major maintenance overhaul base at Denver's Stapleton Airport, but also has submaintenance stations at various locations as needed. Frontier operates as a "passenger-only" airline. The only cargo carried is for the passengers. Flights and maintenance are 99 and 98 percent on schedule respectively (1981). Aircraft operated by Frontier include 49 Boeing 737-200s and 5 DC9-82s (December 1982).

Checks, inspections, and overhaul periods are listed and briefly detailed in Frontier's "Operations Specifications" (FAA Form 1014), figure 28. For example, the service check is normally required where the aircraft has made an overnight stop. The "C" check is made on the 737 Boeing jet aircraft at 250 flight hours (Figures 29 and 30). The "D" check is accomplished at an interval of not more than 22,500 hours[9] and includes an airframe overhaul with total maintenance being performed. The "D" check cannot be postponed without FAA approval. Every part of the aircraft is inspected, repaired, or parts replaced as needed. Tally sheets are reviewed for delay periods, tire changes, actual flight time, recurring problems, etc. A repeat discrepancy report is used and is Frontier's approach toward highlighting critical or repeat problems.

Fixed Base Operator (FBO). This FBO has been introduced to the reader previously, but the role of maintenance in the company was not discussed. The service manager, in addition to maintenance duties, has the responsibility for the library consisting of a multitude of technical orders, service bulletins, and repair manuals for each type and make of aircraft. A full-time

Applications of Maintenance Concepts 101

secretary is required to maintain this reference material plus recording aircraft repairs in accordance with FAA regulations.

It is the service manager's responsibility to maintain and calibrate all of the company-owned test equipment, jack stands, welders, special tools, etc., in the engine shop. It is impossible to overemphasize the importance of maintenance and especially the service manager's position. The aircraft accident potential and related lawsuits resulting from any loss or injury to the public could financially wipe out the company.

The Parts Department's primary function is to support the maintenance operation by having on hand, or readily available, those materials needed to

---

PART D
FRONTIER AIRLINES, INC.

OPERATIONS SPECIFICATIONS

BOEING 737-2A1, 2C0, 2H4, 214
222, 247, and 291
PRATT & WHITNEY JT8D-9A ENGINE
PRATT & WHITNEY JT8D-17 ENGINE

AIRCRAFT MAINTENANCE—PREFACE PAGE
CHECKS, INSPECTIONS, AND OVERHAUL TIME LIMITS

Service Check

A Service Check is performed at not more than 7 calendar days from the last Service Check, Maintenance "A", "B", "C" or "D" Check, and/or when the aircraft RON's at a maintenance station where technicians are on duty, and no higher level maintenance check is being accomplished.

Maintenance "A" Check

A Maintenance "A" Check is performed at not more than 125 hours of aircraft time in service from the last Maintenance "A", "B", "C", or "D" Check.

Maintenance "B" Check

A Maintenance "B" Check is performed at not more than 300 hours of aircraft time in service from the last Maintenance "B", "C", or "D" Check.

Maintenance "C" Check

A Maintenance "C" Check is performed at not more than 2250 hours of aircraft time in service from the last Maintenance "C" or "D" Check.

Mid "D" Check

A Mid "D" Check is performed at an interval of not more than 13000 hours of aircraft time in service or from last Maintenance "D" Check.

Maintenance "D" Check (Plane Overhaul)

A Maintenance "D" Check is performed at an interval of not more than 22500 hours of aircraft time in service.

O.C.—On Condition

"On Condition" items are maintained in continuous airworthiness condition by periodic inspections, checks and services as described in the Frontier Airlines' B-737 Maintenance Manual.

AC—Accuracy Check	FC—Functional Check
BC—Bench Check	HSI—Hot Section Inspection
CC—Calibration Check	MC—Medium Check
EC—Engine Change	ST—Self Test
EO—Engine Overhaul	NDT—Non Destructive Testing
EHM—Engine Heavy Maintenance	RON—Remain Over Night

Effective date *September 24, 1982*

Figure 28. FAA operations specifications form (Courtesy of Frontier Airlines, Inc.)

Figure 29. Boeing 737-200 "C" check, 1983 (Courtesy of Frontier Airlines, Inc.)

get the job done. All aircraft parts, avionics needs, fuel, lubricants, oil, tires, paint, cleaning fluids, etc., and outside contract work is handled via a purchase order system. The manager should be prudent in logistics as well as have a certificated background in aircraft systems. The supply system in the aircraft industry has many outlets and the prices of certain components can vary as much as 100 percent. A reasonable knowledge of what components can be repaired is another asset. Starters, generators, hydraulic pumps, and most cockpit instruments have a core value that can reduce the costs of replacement parts.

The following information represents a reconstruction of an interview with the FBO president's administrative assistant:

> The Avionics (electronics) shop is a highly technical area that requires specially trained people to install and/or repair the myriad of electronic equipment that is found in most general aviation aircraft. It has a function parallel to the company's basic policy—sell aircraft! New radios and navigation equipment can increase the value of used aircraft considerably. An interesting thing about the maintenance of electronic gear: it is normally a total fix condition on what is inoperative or out of tolerance. Customers will bring aircraft in for preventive engine and airframe maintenance but

Applications of Maintenance Concepts

Figure 30. Engine maintenance: "C" check, 1983 (Courtesy of Frontier Airlines, Inc.)

never for electronic work. It was found that heat build-up behind the instrument panel is the source of most malfunctions. The only preventive measures that can be taken is not to allow the aircraft to sit outside in the hot sun.

Before the work atmosphere and trade skills in the shop of preventive maintenance on light aircraft are explored, let's illustrate the value of preventive maintenance on an aircraft. In 1948, Beechcraft produced a general aviation classic called a Bonanza. It was a single-engine, four-passenger monoplane that remains on the production line today. It sold for approximately $9,000 to $10,000 in the early 1950s. A new model, considerably refined, of course, sells for over $90,000. The 1948 model, properly maintained, with present-day radio equipment installed, is worth $16,000 to $20,000! Twice the original purchase price. Preventive maintenance on aircraft is rigidly controlled by FAA regulations. It is quite difficult, however, to determine when repairs exceed the value of the aircraft, indeed, if they ever do. Many owners keep the basic airframe and only update the engines and electronics, and refinish the interior. At some point in time, however, minor maintenance becomes major. Is it time to spend big money for repairs or buy a newer, bigger, faster, latest model machine? For a small airline, it could mean buying a new airplane as capacity and business grows. For a new firm, a used airplane, properly maintained, can mean a lower initial investment. In either case the company is there to assist the customer in that decision. The FBO tries to show the customer that in addition to the safety factor and government regulations, a properly maintained aircraft will hold its resale value.

The major and minor repairs previously discussed seem to describe the two types of jobs most prevalent within the company. Two shop foremen, with the service manager, coordinate the skilled-trade workers on the work to be done in the shop. The 10 general A&P mechanics handle most of the preventive maintenance. A good portion of this work is from customer-initiated work orders. The signed work order is a critical item for work cannot begin until it is authorized. This document is the customer record of parts and labor expended and, of course, is the bill to be paid. Annual inspections, 100-hour inspections, and delivery preparation are all preventive maintenance. The sheet-metal worker, the painter, and the upholstery shop are indispensable to the aircraft sales force. They are available for any job, but trade in aircraft is no different than cars. It is amazing what a paint job and new upholstery can do for the resale of a well-maintained aircraft!

One of the biggest challenges the company has is the salvage of wrecked aircraft. This is almost always a major repair. It requires the concentrated coordination of everyone. From the initial appraisal of damage to customer delivery, it takes on every aspect of maintenance. Questions such as "Do we repair it or buy a new part?" are asked many times a week. New or used parts have to arrive in sequence to facilitate a speedy repair. Every trade specialist gets into the act and occasionally in the way of one another. On one occasion, the maintenance crew laughed at the attempts to reupholster a cockpit at the same time as an engine change, but everything worked great after a tail stand was put in place to stabilize the aircraft. Engine changes can actually be minor as far as workload is concerned. It is the total engine overhaul that is extremely time-consuming. (Many operators prefer to contract out this type of work to specialty shops.)

The crew required for the certified maintenance shop operation must be highly trained and licensed by the FAA. The service manager and two shop foremen require experience, managerial ability, and an additional rating of Airframe Inspector (AI). The FBO is housed in a large aircraft hanger that has adequate space for over 20 airplanes. One aspect the FBO had to consider was the noise environment of being on an airport and its implications on the ability of the workers being able to communicate with regard to a critical operation such as the landing gear retraction tests. The building is heated and cooled in the central offices only. Effective scheduling of jobs has avoided the necessity of workers being out in the sun during the peak heat of summer. During the mild cool of the winters in southern Arizona, work can start in the indoor shop if possible and then progress outside as the work develops. Space heaters are available. The floors are painted and the overall facility is spotless. Daily cleaning is provided by the line crew during the slack periods. Actual building maintenance is minimal; however, one full-time employee acts as the facility housekeeper. His duties

Applications of Maintenance Concepts

also include maintaining those tools, vehicles, machines, air conditioners, etc., that are not controlled by the FAA. If large, extensive aircraft repairs are undertaken, then the whole facility itself must be FAA approved.

In concluding, it is useful to examine the products of the company's maintenance program. FAA regulations require engine and airframe logbooks. Mandatory entries are required as well for routine maintenance. This history of the aircraft is invaluable to the prospective buyer. Its preventive maintenance program provides a hint toward future maintenance problems. In fact, the logbook alone is the best representation of the airplane's worth. A common statement made by aircraft salesmen is: "Show me an aircraft with well maintained logbooks, with flying time properly recorded, and I will show you an airplane that can be sold for top dollar." That aircraft sale will make a profit for the company, and in doing so, one of the primary purposes of the maintenance department is accomplished.

The application of maintenance management concepts in an aviation environment is considerably different from other industries regarding the intensive pressure generated for safety, reliability, and maintainability. The need for management's understanding of the controlling FARs and certification procedures is paramount. Personnel in the maintenance activity are critical to the success of the company, so every effort must be made to understand the requirements for training and certification. Efficient and maximum utilization of apprentices, helpers, mechanics, repairmen, and inspectors must be sought. And, finally, the full range of management tasks, including quality control, administration, labor rates, and much more must be very carefully thought out with decisions made that will make a proper balance of efficiency and goodwill among the company employees.

8

Budgeting, Cost Controls, and Cost Reduction

BUDGETING AND COST CONTROLS are management tools used extensively throughout all departments in a company. A budget is an economic plan representing management's best estimate of expenditures during a specific period. Budgets therefore become statements of anticipated results. When the budget is well prepared, it becomes an effective cost control tool in that reports showing actual expenditures to budgeted allowances provide a basis for corrective action.[1]

To be used as an effective tool of management, the budget must be sensitive to current aviation market conditions. And to communicate effectively its intentions to executive management, the maintenance manager should express the budget in terms of dollars, since this is the most familiar and forceful unit of measure. It also allows top management to make comparisons between departments.

The maintenance department budget becomes one of profit planning, since aircraft maintenance is a charge to someone, either company sales or customer work requirements. As a result, in many aviation activities, maintenance "carries" the company. The use of last year's costs as a budget for current planning is not wise unless the several cost changes that have occurred along the year have been considered. The new budget should recognize improvement factors resulting from improved methods or better

Budgeting, Cost Controls, and Cost Reduction

equipment, cost savings resulting from major rehabilitation or replacement, facility changes completed during the year, and any changes in maintenance personnel staffing and salary structures. As a practical solution for budgeting, maintenance must anticipate a profit as can be seen from the break-even graph shown in Figure 31.

While there are some disadvantages to developing a budget, namely, because they are time-consuming to develop and frequently inaccurate, the advantages far outweigh the problems. Many companies utilize a trend or continuous budget, one that is reviewed and modified, if necessary, quarterly or even monthly. At each review period, it is extended for another period. This method ensures much more accuracy and confidence in the budget concept, allows for new thinking and initiative, improves the quality of planning, and allows the budget to become more flexible and adaptable to changing economic situations. As all other departments must do, maintenance must first analyze and prepare a cost budget—then, as maintenance revenues come in, the break-even point (BEP) becomes of interest.

The logical, systematic approach toward developing the budget includes: making best-guess estimates first, forecasting maintenance activity, preparing the budget proposal, and presenting and supporting the proposal to management.

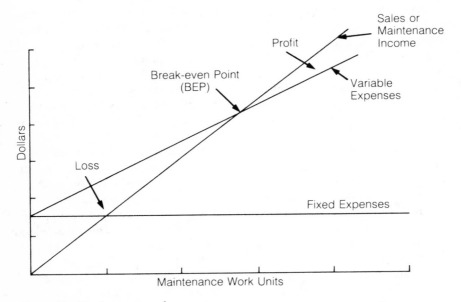

Figure 31. Break-even graph

The budget, in context, must include the following elements: maintenance/replacement of machine and hand tools, shop equipment maintenance, shop facility maintenance, prorated share of overhead and utilities, supervison and administrative support, required purchases, housekeeping and grounds upkeep, aircraft fleet maintenance, and all departmental salaries.

Each new budget begins the function of cost control over the maintenance work to be accomplished (nonrevenue type). During the time evolving, the maintenance manager is looking at: cost expenditures, comparisons with the budget plan, reviewing and analyzing differences, and readjusting, as necessary, the initial budget plan.

If the actual performance greatly exceeds the budget plan, there may be more than one answer that can be identified, such as: poor basic input data, omission of some operating costs, lack of understanding or acceptance of the budget plan initially, and lack of correction on a planned time interval.

Frequently, in an effort to maintain a profit or to minimize losses, maintenance activities and maintenance personnel are cut. This results in a reduced budget that is not sufficient to maintain the facilities. This can become a major problem for the department manager. To correct this situation, the manager has to generate extra paper work showing the multiplier effect of improper budgeting in terms of future costs. Based on current and historical maintenance records, the manager can show excessive equipment downtime, facility deterioration, and other conditions resulting from an overly excesive budget cut. Put in terms of dollars, executive management will not miss the point. "Thus, dollars can be used to tell an organization what it should spend as well as what it will spend for maintenance."[2]

Cost reduction is a way of life in many aviation films, with the larger organizations such as Boeing Airplane Company, General Dynamics/Fort Worth, Ford Motor Company, and others using the technique identified as value analysis, value engineering, or value control. Regardless of the term used to identify the cost reduction method, the methodology itself is the same.

Value analysis (VA) is a unique cost reduction method devised and perfected by Lawrence D. Miles. VA approaches the problem from the basis of the "cost" of a product or process in dollars, to the "function" secured from these dollars. It shows how to analyze accurately the functions of a product, identify what is desired, and place upon each function an appropriate cost. And because every dollar expense of every nature is intended by the purchaser to buy some type of function, this "function-based" technique is applicable to all types of products and services.

Value analysis results in the orderly utilization of alternative materials, newer processes, and abilities of specialized suppliers. It focuses engineering, manufacturing, maintenance, and purchasing attention on one objective—

equivalent or improved performance at lower cost.[3] Value analysis serves all departments of an organization: engineering, manufacturing, maintenance, procurement, marketing, and management. Value analysis is not a substitute for conventional cost-reduction work methods. Rather, it is a totally different procedure for accomplishing considerably cost-savings results.

Inherent in the philosophy of value analysis is full retention for the customer of the esteem and usefulness features of the product. Identifying and removing unnecessary cost, and thus improving value, must be done without reducing in the slightest degree quality, safety, reliability, maintablility, dependability, and the features that the customer wants. No reduction whatever in needed quality is tolerated in the work associated with value analysis. Experience shows that quality is normally increased as the result of developing alternatives for the accomplishment of the use and esteem functions.

The Value Analysis Techniques.[4] The following defines the methodology used in getting better value out of the product, process, or whatever:

Use the value analysis job plan. Value analysis is an organized and systematic search for better value. The VA job plan is what makes it organized and systematic.

Get all the facts. This is a realistic, hard-boiled job. You need facts on costs, inventory, requirements, usage, specs, development, manufacturing, and history. You'll find that in most cases when all the facts are supposed to be in, they're not!

Use information from only the best source. Not just the "normal" source—not the easiest source—but the best source. Get answers about manufacturing from manufacturing, purchasing answers from purchasing, customer needs and wants from no one but the customer.

Define the function. Define the function in two words—a verb and a noun. This forces you to separate functions, to consider them one at a time. It keeps you from considering a part, assembly, or maintenance process as it is, and focuses your creativity on what it does. If a function can be described in two words, it is probably clearly understood. If more words are required, they won't describe the function at all but rather will explain how it is being performed. Write down all the functions that come to mind. Select the function or functions which best describe what you really want, crossing out those that don't apply.

Evaluation by comparison. Value is determined by comparison—no other way. Compare the item with familiar commonplace items that perform similar jobs. What does it look like? What does it make you think of? What else costs about the same? What has similar attributes? These may be in entirely unrelated fields, such as universal hardware or even toys.

Analyze costs. Know the type of cost information that will let you tie specific bits of cost to specific bits of function. The more accurately this association can be made, the better. Know the shop cost (as far as possible) of: each component; each manufacturing process applied to a component; each step of maintenance, etc.; each tolerance, special finish, etc.; and each inspection. Consider all recurring costs of: raw material; direct labor; overhead; inspection; purchased parts and material; rejection, rework and retest, scrap; and maintenance. Consider nonrecurring costs (portions that have not been expended): engineering, tooling, planning. Know cost per unit. Consider units to be manufactured. Know shipping and packing costs. Consider total cost. Are you influencing maintenance costs, for example?

Do creative thinking (brainstorming). Turn off the judicial part of your mind. Put your imagination to work. Apply creativity to the functional definitions, not to products. What else will do the job? In what new and unique ways can the function be performed? Use both individual and group creativity—brainstorm! SUSPEND JUDGMENT! Bring out as many possible alternatives as you can, regardless of how impractical they seem at first. Save judgment until later.

Evaluate the function. Define each function. Identify basic and secondary functions. Evaluate each function by comparison. Ask—and answer—five basic questions: (1) What is the item? (2) What does it do? (3) What does it cost? (4) What else will do the job? (5) What does that cost?

Blast, create, and then refine. It's more effective to take a simple concept and build up to meet all needed requirements than it is to take a complex idea and try to identify and eliminate all the elements of unnecessary cost. *Blast.* Blast down to basic functions only—strip them bare by temporarily disregarding all other parts of the product or by accepting them as is, and by temporarily disregarding secondary functions. *Create.* Do creative thinking to develop alternate ways and means of accomplishing the oversimplified concepts revealed by the blasting. These should be the simplest and least expensive ways you can think of, even if you know full well they are not entirely satisfactory. *Refine.* The refining comes about when you start to adapt the ideas so they are satisfactory. Other refinements will appear when the secondary functions are reviewed. Continue creativity. Knowingly add increments of function along with additional increments of cost until the refined product fully accomplishes the total function. Then STOP.

Identify and overcome roadblocks. Anticipate roadblocks. Evaluate whether roadblocks are real or imaginary. Roadblocks can be things; they can be situations; they can be people. Most roadblocks are the result of habit—or attitude. Look out for self-imposed roadblocks. Identify the specific roadblock. Define it as clearly and sharply as possible. Then set about overcoming it.

Challenge requirements. Challenge requirements and specifications for such things as finishes, tolerances, stress, tests, inspection and delivery schedules. These usually cost a lot of money. Finding an unnecessary, unwanted, or obsolete requirement will pay quick dividends. Here are some reasons why requirements should be challenged, reasons why they may be necessary: changing conditions—what was true then may not be true now; poor communications—midunderstanding as to actual requirements; lack of time when original decisions were made; product patterned after a prototype; lack of cost knowledge—not knowing the specific cost of the requirements; unknown technical requirements—stress, heat, radiation and pride of creativity on the part of whoever made the specification. At one time specifications and requirements were sacred. No one dared question them. It was assumed that the customer or whoever made the specification knew that it would produce what he or she wanted. Value work has shown over and over that specificatons and so-called requirements are only methods used to obtain the functions that represent real value to the user. If anything is sacred, it's the customer-desired functions and features—not the specifications.

Use specialty vendors. Draw on the almost unlimited resources of industry. Take advantage of the know-how of people who make a business of producing the product, process, or service you want—or one like the one you want. Use specialty products and processes. Look for vendors who specialize in manufacturing, processes, testing, design, stamping, molding, etc. Use job shops. Find large and small quantity producers. Look for newly developed processes, materials, tools, and machines. Ask Purchasing and Vendor Research to recommend specialty vendors. Avoid expensive "do-it-yourself" approaches. Don't reinvent the wheel.

Use the company's specialists. Most companies have people and departments whose sole purpose it is to give specialized service. Take advantage of what's already there. Don't hesitate to seek out and consult company experts.

Use standards. Search diligently for a standard. Standard products are inherently less costly than "special" or "one-time use" parts. Can the product be designed to accept a standard part rather than designing a "special" part to fit the product?

Use good human relations. Human beings dislike change. Value work nearly always involves a change of some sort. People may be suspicious of your motives, and they probably won't want to accept your ideas. Treat people with the dignity and consideration they have the right to expect. Avoid any appearance of being critical. Give willing and enthusiastic credit to others. Try to make them a part of the opportunity to get better value.

Get new information. Bring new information into each functional area. Find what new products, materials, or techniques have been developed since the last time someone studied the item. Find what changes in requirements have been made.

Get a dollar sign on every given tolerance. Tolerances cost money. So do finishes. Generally speaking, a tight tolerance costs more than a loose one. If possible, determine and assign a dollar value against inspection, fabrication, and test tolerances. The tolerance (or finish) spelled out on a drawing might force a milling machine operation when a cast or welded part will do the job. In other words, close tolerances may cost dearly but contribute no value. Be sure those called for are needed and not the result of habit. If they don't stand up under investigation, they aren't needed. Define and evaluate the function of the tolerance or finish.

Avoid generalities. "It always costs more to machine a part than it does to fabricate it." "It's not practical to buy these because we can't get good quality." Recognize these generalities as a force that maintains the status quo. The first step must be to eliminate to generality in order to come to grips with the specific problem. What does this specific item cost? Can we buy, cast, fabricate, machine, eliminate, etc., this specific item in this specific case?

Use your own judgment. How much of your time do you put in doing things exactly as you think they should be done? Someone has said that value analysis is "organized common sense." If you're going to get value, it's vital to do what makes the best sense. Any deviation from answers that make the best sense will result in either lower performance or lower value.

Use the criterion, "Would I Spend My Money This Way?" When average individuals spend their own money, they are governed by the following typical conditions: They have a limited amount to spend. They try to get the most use and appearance they can for their money. They expect to get these functions within reasonable limits in return for their money. If they don't, they make appropriate changes—either then or the next time they get ready to spend their money. They know they can't get reasonable value in exchange for their resources unless they have value alternatives clearly set up and use all the information they can get hold of to make decisions. Before they spend their money, they will have a good idea of the relative use values, the relative esteem values, and their relative costs.

It has been stated that VA has applications in all types of industries and in all facets of management from engineering to maintenance to administration. Parts, purchases, records, inventory quantities—all are subject to value analysis and cost reduction.

For example, when analyzing quantities to be stocked, the maintenance activity should be aware that there may be a better material or process that will reduce the total cost and assure a higher profit for the company. Instead of stocking twenty or more different lubricants, it may be possible to consolidate the types and stock something less than ten. In another area, it may be possible to use treated paper filters instead of throwaway fiberglass filters. Where conventional bearing seal failures are high, consider the installation of mechanical seals. Replacing single-element electrical fuses with dual-element fuses often prevents costly power outages. These are only a few situations in analyzing inventory quantities alone where savings can be generated through value analysis.[5]

Budgeting, cost controls, and value analysis are an inherent part of every sizeable aviation activity. These functions are becoming more prominent due to high interest factors, scarcity of cash, reversed cash flow, and the necessity to cut back on operations—all operations, maintenance included.

Good, accurate budgeting and the ability to hold to the budget plan is paramount in today's requirement for the maintenance manager. A defined way of assistance in holding to the budget plan includes the incorporation of value analysis into the economic system. Fortunately for the maintenance manager, there is usually no extra cost involved in utilizing VA techniques as a starting point for personnel. Of material interest to the manager is the fact that all maintenance personnel can get involved in value analysis applications (see Appendix C, which includes VA definitions).

9

Training and Professional Development in Aviation Maintenance

AVIATION SCHOOLS ARE A MUST to the progress of the aviation industry. They must provide the replacement personnel for those that retire and the great amount of personnel it takes for the industry to expand. The mechanics have to be trained and certified in accordance with the appropriate Federal Aviation Regulations. Aviation schools do this and most go a step further in teaching safety beyond what is called for by the Federal Aviation Administration. These schools are certified under FAR Parts 141, 143, and 147 to train the students to become Aviation Maintenance Technicians (Airframe and Powerplant Certified Mechanics), Avionics Technicians, Ground Instructors, Pilots, and Air Traffic Control Tower Operators. Aircraft manufacturers, aircraft parts manufacturers, airplanes, fixed base operators, and repair stations look to the schools for the trained personnel to sustain, improve, and enlarge their operations.

The general prerequisite for admission to all schools in the aviation field is for each student to be at least seventeen years of age and have graduated from an accredited high school. Students may enter by examination if they are eighteen years or older and have completed one of the following: (1) taken the General Education Development test and made a satisfactory high school equivalency score, (2) completed a composite standard score on the American College Testing Program or equivalent which would place them among the upper three-fourths of high school seniors based on the twelfth grade national norms, or (3) have a satisfactory score on the school entrance examination.[1]

Admission to flight school and the issuance of a private pilot certificate requires the student to be at least seventeen years old and have a third-class medical certificate issued within the last twenty-four months. To be issued a commercial pilot certificate the student must be eighteen years old and have a second-class medical certificate.

International students are admitted to some aviation schools if they meet certain requirements listed in FAR Part 65 and Part 147. Students from countries with a native language other than English must establish their English proficiency. This can be done by a score of five hundred or higher or an official TOEFL (Test of English as a Foreign Language) or by earning thirty semester credits at an accredited college or university where English is the primary language.[2] Some technical schools and all colleges that offer aviation courses also offer English to foreign students. In this way international students can enter many American schools that offer aviation training.

The airframe and powerplant licenses required by the FAA for all aircraft mechanics have certainly led to an avoidance of many of the problems that plague many of the other production and service-related industries. An aircraft mechanic by the very nature of the work cannot hang out a shingle and thereby declare himself or herself duly qualified. He or she must be a highly trained individual and possess the sense of responsibility that comes from knowing people's lives are at stake. Ninety-five percent of all A&P mechanics are now trained in a formal environment such as the military, trade schools, or part of a college curriculum. It is possible for an individual to obtain these credentials without ever having attended a formal school; however, the time and effort in doing so is considerable. The A&P license is a mandatory requirement to be considered for employment at any aviation firm. Graduates from a school program are considered apprentices in the industry although they are far better qualified than any other tradesmen. (An example of an aviation "trade" school is introduced subsequently in this chapter.) For individuals not pursuing a college education, the A&P apprentice mechanic has one of the highest paid starting salaries in the United States. A newly hired individual at a company will work with a senior mechanic on all minor and major repairs until such time that the service manager is convinced of the individual's ability to handle the job.

Each applicant for a mechanic certificate or rating must pass an oral and practical test on the rating he or she seeks. The tests cover the applicant's basic skill in performing practical projects on the subjects covered by the written test for that rating. An applicant for a powerplant rating must show his or her ability to make satisfactory minor repairs to, and minor alterations of, propellers.

The written tests required for the A&P license can be taken at FAA flight standards district offices. Before students can test, they must meet the experience requirements for the rating they are after (A license or P license or both). Once eligibility is confirmed by the FAA, students may test.[3]

There are three separate tests for an A&P license: (1) Aviation Mechanic General Test, (2) Aviation Mechanic Airframe Test, and (3) Aviation Mechanic Powerplant Test. Two hours are required for the General Test and four hours each for the Airframe and Powerplant tests.

The student must pass the General Test and either the Airframe or Powerplant tests within a 24-month period in order to be eligible for a mechanic certificate and rating. The tests are made up of multiple choice questions, and there are no trick questions.[4] After finishing the test, student answer sheets are forwarded by the FAA to a central location for grading.[5] The minimum passing score is 70 percent. Test scores will be mailed back to the individual normally within five working days. These scores should be kept safe as it is imperative that they be presented for Oral and Practical examinations.

The oral and practical tests are usually the final step in getting the requested rating. An oral and practical must be taken for each rating. Oral and practical tests are given by FAA flight standards inspectors or by FAA designated mechanic examiners (DME). If an FAA inspector gives the test, the student must furnish or arrange for tools, materials, supplies, and a proper facility. If a DME gives the tests, the examiner will furnish the facility and will usually arrange to have the required tools and materials available.[6] The names and addresses of the FAA DME in each district can be obtained fom the FAA flight standards district office that serves the area or from Advisory Circular 183-32A, FAA Designated Maintenance Technician Examiner Directory.

Oral tests may be given along with the practicals in the form of questions about the tasks being performed, or these can be taken separately. The oral test questions cover the same subjects as the written tests but they're intended to show how well the student makes use of his or her knowledge right on the spot.[7] There are three types of oral test questions: (1) questions related to practical projects, (2) questions not related to practical projects, and (3) questions to determine if more projects need to be assigned.[8]

The practical tests are a different project to test the student's mechanical ability and his or her ability to organize the work to be done. The instructor will pick projects that utilize, as much as possible, tasks with which the student is familiar. Orals and practicals are graded as soon as they are finished. The students are then informed of their grades. If any part of the test is failed, the examiner will issue a disapproval notice showing the titles of the subjects failed. If a person fails a written, oral, or practical test, he or she must wait 30 days to retest. The student can test before 30 days only if he or she gets a signed statement from a certificated mechanic holding the rating the student is after, certifying he or she has given the student additional instruction in the areas failed.

Each student is required to complete 1,900 hours of instruction at any technical school. Depending on the institution, some run five week blocks, some by semester. They are programmed up to two years of study.

A certificate is earned by completing all courses successfully. Additional general education courses are required and most offer an associate of science degree.[9] The advantage in attending a technical school is being able to take the oral and practical examinations during the final phases of the curriculum, before taking the written exams.[10]

The aviation maintenance technician program is designed to teach the technical skills required to exercise the responsibilities of an airframe and powerplant technician. Successful completion qualifies the graduate to take the written, oral, and practical tests administered by the Federal Aviation Administration for the airframe and powerplant certification.

The students start the course by taking general subjects for approximately five hundred fifty hours give or take fifty hours depending on the school. The subjects may vary some from school to school but the course objectives are basically the same. The following list of general courses taught by Spartan School of Aeronautics is an example of what is normally required: (1) basic mathematics, (2) basic physics, (3) basic mechanics, (4) basic electricity, (5) aircraft weight and balance, (6) maintenance, materials, and processes, and (7) drafting.[11] These general courses give the knowledge and skill to continue on in the field of airframe and power plant technology.

The powerplant technicians' course is designed to teach the technical skills required to exercise the responsibility as a powerplant mechanic. The course covers every aspect of the engine and propeller system. The course covers the general subjects listed in the preceding paragraph plus the special subjects that are required by the Federal Aviation Administration. The special courses are designed to make one familiar with the reciprocating and turbine engines and the propellers used on the specific engine. The student will have the opportunity to become familiar with the teardown and overhaul procedures of the engine. He or she will also gain valuable experience in troubleshooting and repairing the electrical and fuel systems.

After successfully completing this course of twelve to eighteen months, the student will qualify to take the written, oral, and practical test with the Federal Aviation Administration for the Mechanic's Certificate with the powerplant rating. This will allow the certificated mechanic to approve and return to service a powerplant or propeller or any related part or appliance, after he or she has performed, supervised, or inspected its maintenance or alteration (excluding major repairs and alterations). In addition, he or she may perform the 100-hour inspection required by FAR Part 91 on powerplants or propellers and approve and return such equipment to service.[12]

The airframe technician course is designed to offer the student skills and knowledge in repairing of the aircraft minus the engine, propellers, and radios. Upon the successful completion of the general and special courses, the student can take the Federal Aviation Administration's written, oral, and practical tests which would qualify the graduate for the Mechanic's Certificate with the airframe rating.

After completing the genral subjects listed previously, the student will enter into the special courses which provide him or her with the skills to perform inspections and maintenance on the airframe. These include the repair of electrical systems, sheet-metal, hydraulic systems, welding, fabric and wood structures, and auxiliary systems.

After the student has graduated and passed the Federal Aviation Administration test and has been certified as an airframe mechanic, he or she may approve and return to service an airframe or any related part or appliance, when he or she has performed, supervised, or inspected its maintenance or alteration (excluding major repairs and major alterations). In addition, he or she may perform the 100-hour inspection required by FAR Part 91 on the airframe or related parts and approve and return to service.[13] During the twelve to eighteen months of this program the student has gained much knowledge and many skills that are applicable to maintenance industries other than aviation.

The avionics technician program requires a highly motivated individual with an excellent education in math as a prerequisite for this course in any school. The students should have a basic knowledge of circuits but this is not required in all schools that teach avionics.

A few schools teach the avionics from the bottom up, beginning with a general knowledge of electrical and electronic fundamentals, including basic circuitry. These courses usually last a period of eighteen months.

The students will become familiar with standard circuits encompassing the principles of amplification, detection, rectification, modulation, wiring wave shaping and servo-mechanisms as they progress through the course.

By the time they have completed the eighteen-month course, they will have become efficient with the application of the primary test equipment, such as vacuum tube testers, multimeters, oscilloscopes, and signal generators that are used to test electronic circuits.

When the students have completed the course, they will be given the opportunity to test and be certified by the Federal Communication Commission as Avionics Class One Certificate holders. These technicians will be able to work on the radios or electronics of any FAA certified aircraft.

Colorado Aero Tech[14] represents the caliber of vocational school requirement by the FAA to assure that proper training in all facets of aviation maintenance and safety has been accomplished for all students seeking an A&P license.

Colorado Aero Tech, Inc., an affiliate of Frontier Airlines, originated in 1958 as the Kensair Corporation, at Peterson Field, Colorado Springs, Colorado. At this location, Kensair consisted of a full base operation including an FAA-approved flight school, FAA-approved repair station, and Piper sales dealership with storage and refueling facilities.

As the company continued to grow, it purchased in 1961 a lease on hangar and ramp facilities at Jefferson County Airport, Broomfield, Colorado. These facilities included a hangar building of 33,000 square feet, 17 tee-hangars, and 4 acres of ramp area.

In May 1963, Kensair Corporation was issued Air Agency Certificate No. WE-03-17 by the FAA with approval of the following ratings: Basic Ground School, Advanced Ground School, Primary Flying School—Airplanes, Commercial Flying School—Airplanes, Instrument Flying School, Flight Instructor School, and Commercial Flying School—Helicopters. Kensair added a technical training Airframe and Powerplant course on June 1, 1965. This course was issued an Airframe and Powerplant School Certificate No. 4614 by the FAA. Colorado Aero Tech, Inc., Division of Kensair Corp., became the official name of the A&P school on November 20, 1966.

In August 1967, Colorado Aero Tech added a brick building constructed specifically as an educational institution with classrooms, administrative areas, shop areas, and reciprocating engine test cells. In 1970, a large, all-metal hangar was added to the original brick building and in 1972, a separate test cell housing both recip and turbine engines was purchased. By 1976, the school was operating at a full capacity of approximately 300 students and in 1978 permission was granted by the FAA to increase the number of students from 300 to 325.

In the winter of 1979–80 a decision was made to add additional facilities to the school and to introduce at that time a night program which was to be a duplicate of the day program in terms of hours, curricula, etc., which would then allow the school to grow to a maximum capacity of 650 students. An additional 13,000 square foot hangar and classroom building was built east of the main hangar and both day and night school students began officially using the building in October of 1980 (Figure 32 shows a student-assembled CJ-805 GE turbofan engine).

In December 1982, Colorado Aero Tech was sold to Frontier Services

Figure 32. Assembled General Electric CJ-805 turbofan engine, 1983 (Courtesy of Colorado Aero Tech, Inc.)

Company of Denver, Colorado. Frontier Services is a wholly owned subsidiary of Frontier Holdings, Inc., which also owns Frontier Airlines. Colorado Aero Tech was purchased by Frontier Services Company to be the cornerstone in what is hoped to become eventually an eight- to ten-school technical training division with schools located in various parts of the country.

Another type of training/educational facility, the four-year aviation curriculum, is illustrated by Southern Illinois University's (SIU's) several aviation programs. The SIU aviation program is so complete that the university was awarded the FAA's top curriculum rating in 1983 (comprehensive rating). The school offers complete A&P training and flight training instruction. In addition, SIU offers an Associate Degree in Aviation Flight, a Bachelor of Science Degree in Technical Careers with a major in Aviation Management, and an Airway Science Curriculum.

The associate degree program includes flight courses through the commercial pilot certificate and the instrument rating. The flight instructor certificate course and the multiengine course also are included. Flight theory courses supplement and complement flight courses. General education and basic science courses supplement aviation-related technical courses to enhance the student's professional value to the aviation industry.

The Bachelor of Science Degree in Technical Careers with a major in Aviation Management or Career Development provides graduates of the associate degree program in Aviation Flight an excellent opportunity to complete requirements for the bachelor's degree in two additional years. The Aviation Management major offers a structured curriculum designed to meet students' needs in entering the aviation industry, while the Career Development major allows the student more opportunity to design an individualized program to meet more specific career goals.

SIU is also an approved FAA Airway Science Curriculum school. The curriculum emphasizes courses in aviation, science, and technology, mathematics, management, and general education programs. Applicants for FAA positions as air traffic controller, electronics technician, aviation safety inspector, and computer specialist who enter through the Airway Science Curriculum will be ranked and selected from a separate register parallel to those currently in use.

Aviation schools range from the small technical school with some aviation courses to the large university that teaches aviation subjects within its curriculum. The small technical school might teach only flying or avionics but with the strict FAA rules that all schools must abide by, the student can count on receiving all the material required to pass the FAA test. If the student decides to be a mechanic, electrician, or pilot, there is a school close by to give the instruction needed to pass the FAA test and become certified to work in that chosen field. The airlines and general aviation would be hard put to train enough mechanics or pilots to keep aircraft flying, so schools become more important as time passes.

10

Safety and Maintenance

WHEN THE COMMERCIAL JET AGE BEGAN 25 years ago, there were people who claimed the new high performance aircraft were too radical a departure from normal flight.[1] They said jets were too much plane for any man to maintain and handle safely, would not fit into air traffic control, would be operating in the unknown environment of higher altitudes, etc. In brief, pessimists predicted that the airlines' fatality rate would climb higher than the new jets themselves.

By every statistical yardstick that can be applied to safety performance, these predictions were wrong. On the basis of fatalities per 100 million passenger miles flown, the most common standard of safety measurement, the airline fatality rate has declined steadily since the government began keeping accurate records in 1937. United States carriers compiled the rates per 100 million passenger miles compared with automobile travel shown in Table 8. In 1959 and 1960, the first full two years of worldwide jet operations, the new planes averaged one fatal accident for every 150,000 hours of flight. By 1977, United States airlines were flying nearly three million hours per accident.

Air travel is safe, largely because safety is not only the highest priority of the industry, but is almost a religion. This is a statement substantiated by facts that go beyond statistics. Many foreign carriers send their flight crews to

TABLE 8
PASSENGER FATALITIES PER 100 MILLION PASSENGER MILES (FIVE-YEAR AVERAGES)

	Air	Auto
1942–46	2.08	2.74
1947–51	1.65	2.20
1952–56	.43	2.74
1957–61	.51	2.32
1962–66	.21	2.36
1967–71	.18	2.20
1972–76	.09	1.56
1976–80	.056	

Source: National Transportation Safety Board (NTSB), *News Digest* (1982).

United States airline training centers to learn American methods and techniques. The overwhelming majority of delays for mechanical reasons are based on safety considerations. No aircraft moves from a gate unless all components directly or even indirectly associated with safe operations are functioning perfectly.

There is no compromise with man and machine. An airline pilot's training is continuous, whether a pilot has 3,000 hours in his log book or 30,000 hours. Flying ability and command capability are tested twice a year, and refresher courses are required annually. The same rigid standards are applied to the machine, from drawing board to operating schedules. The wings of a modern jetliner are 50 percent stronger than required by federal regulations. In one structural test, a wing was deliberately deflected some 30 feet upwards out of its normal level without popping a rivet, and the flight tests for this same plane lasted a full year at a cost of $28 million.

Just as training never ceases for a pilot or flight attendant, neither does the inspection, testing, maintenance of an airliner. A jet receives five manhours of maintenance for every hour of flight, at an annual cost of nearly $1 million per aircraft. United States carriers average 23 mechanics per plan in their fleets and spend approximately $2 billion a year on maintenance.

Yet that one-in-a-million chance accident still occurs. Consider the tragic accident that occurred on May 25, 1979: an American Airlines DC-10 crashed shortly after it began a takeoff roll for O'Hare International Airport on a flight bound for Los Angeles.[2] As the plane's nose lifted off the ground, the left engine and pylon tore away, traveled over the wing, and fell to the ground. The jetliner climbed to about 325 feet, rolled to the left, and plunged into a field beyond the edge of the runway. Two hundred and seventy-three persons died in the accident, the worst single aircraft air disaster in United States history.

The National Transportation Safety Board determined that the probably cause of this accident was the asymmetrical stall and the ensuing roll of the aircraft because of the uncommanded retraction of the left wing outboard leading edge slats and the loss of stall warning and slat disagreement indication systems resulting from maintenance-induced damage leading to the separation of the number one engine and pylon assembly at a critical point during takeoff. The separation resulted from damage by improper maintenance procedures, which led to failure of the pylon structure.

Two days after the crash, the safety board issued a safety recommendation calling for the immediate inspection of all pylon attach points on all DC-10 aircraft. A week later, the safety board issued two more recommendations based on its observations of the maintenance procedures used by some carriers to remove and reinstall the pylon and engine as a unit on the DC-10. One of these recommendations urged the Federal Aviation Administration to require an immediate inspection of all DC-10 aircraft in which an engine pylon assembly had been removed in order to check for damage to the pylon aft bulkhead, including its forward flange and the attaching spar web and fasteners. The recommendation also urged the agency to require the removal of any sealant which could hide a crack in the flange area and employ eddy-current or other approved inspection techniques to ensure detection of any possible damage.

The basic cause was incorrect maintenance procedures, resulting in a flight accident that originated on the ground. Unsafe maintenance and ground operations are causes of flight accidents and it only takes one accident (the DC-10 accident, for example) to alter materially an aviation safety trend.

The role of the FAA in the aviation industry was introduced earlier in the text. While much is known about its responsibility and authority in mandating maintenance requirements, not as much has been written about its parallel responsibilities for the incorporation of safety in all aircraft. It must be remembered that the FAA's mission is directed primarily to aviation safety and to the establishment and enforcement of safety standards applicable to every aspect of civil air transportation. Thus, the direction and overview the FAA exercises over all aspects of aviation maintenance reflects on its mission for safety in the skies and on the ground.

Air carriers, commuters, air taxis, general aviation, owners, and operators are all aware of their responsibilities to the FAA and the flying public for the need for well maintained, reliable, and safe aircraft. Poor maintenance practice has contributed to numerous aircraft accidents and hundreds of fatalities. So it is to be expected that the FAA is always concerned about the maintenance procedures and workmanship put into the civil aviation fleet.

An example of the FAA's concern for good maintenance practices and safety is its issuance of Airworthiness Directives (ADs). In the cases of aircraft, engines, propellers, or parts that it has certificated but later found to be suffering from an unsafe condition, it issues airworthiness directives (ADs), which requires mandatory correction of the deficiency. The ADs are made a matter of record and issued to the public in two separate summaries or volumes: one for small aircraft (that is, an aircraft with an actual certificated takeoff weight of 12,500 pounds or less); the other for large aircraft (aircraft with a certificated takeoff weight above that figure). Each volume consists of two books. Book 1 contains ADs issued prior to December 31, 1970. Book 2 contains those issued since that date. The AD system makes aircraft owners, operators, and maintenance personnel aware of unsafe condition(s) and tells those involved what corrections and/or limitations are to be imposed.

ADs may be issued in any one of the three categories: Emergency—normally immediate corrective action is required; Immediate Adopted Rule—issued when a condition is hazardous enough to constitute urgent action (30 days maximum requirement); and Notice of Proposed Rulemaking (NPRM)—used when no immediate danger or major hazard exists (normally 60 days or more for compliance). Also involved in aircraft airworthiness and, in particular, aircraft accidents, is the National Transportation Safety Board (NTSB).

The National Transportation Safety Board (NTSB) is an independent federal agency that serves as the overseer of United States transportation safety. Its responsibilities are intermodal in scope and include railroad, highway, pipeline, marine, and civil aviation transportation. The mission of the safety board is to improve transportation safety. This is done primarily by determining the probable causes of accidents through direct investigations and public hearings and secondarily through staff review and an analysis of accident information, through evaluations of operations, effectiveness, and performance of other agencies, through special studies and safety investigations, and through published recommendations and reports to Congress.

Related to the aviation field, under the Independent Safety Board Act (effective January 1975), the board has the authority to investigate, determine the facts, conditions, and circumstances, and determine the cause or probable cause of civil aircraft accidents. In air transportation, the board determines the cause of all civil aircraft accidents that occur in the United States, a function that cannot be delegated to any other department or agency. The board investigates all air carrier accidents, most fatal light plane accidents, and other selected accidents. The board has, however, temporarily authorized the FAA to investigate most nonfatal light plane accidents and helicopter accidents but retains the statutory duty to determine probable cause in each case.

General aviation fatal accidents—non–air carrier—are usually handled by the board's field office nearest the crash site. One investigator is in charge of each case, and if a specialist in any category of skills is needed, the field office requests assistance from Washington. Almost always the probable cause of general aviation accidents is decided by the board without a public hearing, based on an analysis of the investigator's findings.

Reports on the causes of all civil aircraft accidents are released publicly; records of major air carrier cases and other selected accidents are periodically issued as individual, narrative reports; in addition, all accident reports appear in synoptic, computer-printout volumes issued several times each year. Safety recommendations are based on information developed from board investigative findings and special studies[3] and are aimed at preventing accidents and correcting unsafe conditions in transportation. The board's recommendations are not mandatory; however, as required by the act, all are made public and they have a high rate of acceptance.

A considerable amount of time and money is spent by the FAA, NTSB, NASA (National Aeronautics and Space Administration), and the air carriers to find ways to improve aviation product safety, maintenance concepts, flight control systems, and many other aspects that could affect the safety of air crews and passengers. One interesting safety experiment involved an actual programmed jet crash to test blast prevention.

A 20-year old jet that has made 53,470 safe landings will be smashed into the ground in one of the biggest aviation safety experiments in history, federal officials say.

The experiment, scheduled for 1984, is to determine if an additive to jet kerosene can prevent explosions on impact.

The Boeing 707-720, assembled in Renton, Wash., has logged 20,000 hours in the air. The four-engine plane, called N-23, has been used by the Federal Aviation Administration to fly research missions and to fly in conditions similar to those in which fatal crashes occurred.

The more than 52,000 landings are the most any Boeing jetliner has ever made.

During its last flight, according to the National Aeronautics and Space Administration, crewmen will make a final check of N-23's new seats, make sure its tanks are filled with fuel and send it barreling down the runway by remote control. The plane will climb, then be directed back over the Mojave Desert where it will smash into the earth.

NASA and the FAA are hoping there will be no fire and that the $10 million test will be a success.

"It will probably be the agency's single biggest aviation safety experiment," said Mike Benson, an FAA spokesman in Atlantic City.

"One in five people killed in civil aircraft accidents die from fire and toxic gases rather than from impact injuries," said Russ Barber, project supervisor at Edwards Air Force Base.

Safety and Maintenance

The key to the experiment is whether a polymeric anti-misting additive to jet kerosene developed in Britain will work. It is the fuel mist that causes the explosions and fires that often incinerate airplanes during crash landings.[4]

Regrettably, the test when conducted was a major failure; but the FAA is apparently going to require the use of the apparently ineffective fuel additive anyhow—at great cost to the airlines and ultimately to the flying public.

Unfortunately for the aviation industry and especially the air carriers, there is no more highly publicized area today than aircraft accidents and their attendant statistics. Aircraft accidents do not just happen—many tangible and intangible circumstances are involved. Many aircraft accidents result from some maintenance failure that occurred on the ground.[5] In addition, aircraft ground accidents occur frequently, but, fortunately, these are rarely serious enough to make newspaper headlines. However, depending on the circumstances, an occasional ground accident is publicized. For example, an accident occurred involving an aircraft worker at China Lake, California.[6] During maintenance work on a plane at the Naval Weapons Center, an aircraft electrician was killed when he was sucked into the intake of a high-powered jet.

Now take a look at a major airline (American Airlines) to see how their safety program is laid on and to review, briefly, how they statistically record each accident, both flight and ground phase. To start with, their safety program is headed by a group of ex-pilots with extensive aviation safety experience. This group is headquartered at the corporate level and is responsible for all facets of AA's safety program, including reporting, training, corrective actions, and other safety concerns. The group's well-defined accident prevention checklist

> is designed to assist station general managers/city Vice Presidents and local functional management in an overall self-evaluation of the effectiveness of the accident prevention program at their respective stations. Much of the material covers areas that are evaluated during the safety division's periodic station safety surveys. . . .
>
> A basic premise of sound accident prevention technique is to learn from past experience. Consequently, many of the items contained in this checklist were derived from actual contributory cause factors in accidents experienced by our airline and other airlines. By analyzing past accidents, it was determined that the majority were caused by human error; as a result, we have placed an emphasis on unsafe acts and practices through the checklist.
>
> The checklist is organized into six sections for each reference. . . .
>
> Part I. *Safety Administration*, is based on establishing safety principles that were found to be effective in countless, successful programs utilized by diverse industries. These principles were then adapted to fit the airline setting. . . .

Part II. *Aircraft Arrivals and Departures,* . . .

Part III. *General Condition of the Ramp and Ground Equipment*, contains a checklist of potentially "unsafe conditions."

Part IV. *Freight Services,* . . .

Part V. *Automotive/Facilities Maintenance,* . . .

Part VI. *Aircraft Maintenance Hangar Facility*, covers specific safety items inherent in these functional areas.[7]

American Airlines' safety division develops a comprehensive annual safety report, one copy of which is given to the president of the airline accompanied by an oral presentation from the senior director of safety. This report contains an initial summary, followed by accident costs covering two years—the current year and previous year—so comparisons can be made. Accident costs are broken down into: aircraft damage, FOD (Foreign Object Damage)—aircraft engines and tires, ground equipment damage, total damage costs, and employee injuries—total compensation costs.

The immediately following lettered paragraphs (A through D) are derived from the first page of the "AA Accident Prevention Checklist," which was designed to assist station general managers/city vice presidents and local functional management in an overall self-evaluation of the effectiveness of the accident prevention programs at their respective stations. Much of the material in the checklist covers areas that are evaluated during the safety division's periodic station safety surveys. In a sense this checklist can be a tool for anticipating problems and correcting them before an outside audit is conducted.

A. Pushout and Tow In: Is the "Mechanic's Ground Handling Checklist" followed? Is the correct checklist used for the type aircraft involved? Are hangar doors completely opened prior to any aircraft movement into or out of the hangar? Are wing walker(s) positioned at the wing tip area or where clearances are critical for adequate visibility and not in close proximity to the main landing gear?

B. Parking of Aircraft (Hangar): Is the aircraft nose gear chocked? Is the aircraft grounded? Are the cabin door barrier straps used when cabin doors are left open and no truck is positioned at the door?

C. Work Practices: Do personnel stand on elevated workstands without using guardrails? (OSHA 1910-29) Are the guardrails adequate, i.e. not makeshift rope-type guardrails? (OSHA 1910-23) Are personnel in compliance with the procedure prohibiting standing on the top step of a ladder? (OSHA 1910–25) Are workstands, ladders or safety harnesses used as necessary when working on high surfaces such as aircraft structures? (OSHA 1910.21-.30) Are personnel using unauthorized equipment as a substitute for a workstand such as a tractor? (OSHA 1910.66b) Are personnel using appropriate protective equipment for the task involved (safety glasses, face

Safety and Maintenance

masks, ear muffs, gloves, aprons, boots, etc.)? (OSHA 1910.132–.136) Are tools and other items accounted for after the task is completed? Do employees leave tools or other items in any part of the aircraft, particularly the engine inlet, while working on the aircraft? During transfer of flammable and combustible liquids from bulk storage containers to portable vessels, are the containers bonded to neutralize static discharge? (OSHA 1910.107)

D. Condition of Hangar: Is the hangar clean and well-organized? (OSHA 1910.22a) Any potential FOD, such as metal objects? (AA Reg. 55-16) Any hydraulic, oil or water accumulations presenting slipping hazards? (OSHA 1910-22a) Are the ground power cable heads stored properly? (GTM Sec. 3) Are the fire extinguisher requirements met for hangar facilities? (AA Reg. 55-3 and Part III, G) Is an unobstructed access to the fire extinguisher maintained at all times? (OSHA 1910.157a) Are exits maintained free of obstructions? (OSHA 1910.37)

Continuing in the AA safety division report is a detailed breakdown of the types of accidents causing aircraft damage. These breakdowns come under a ground phase heading covering such areas as: aircraft, mechanical failure, and failure to disconnect. The flight phase data includes: objects on runway, bird strikes, engine fire, landings/takeoffs, mechanical failure, weather, and miscellaneous. The American Airlines' report includes a detailed description of all major damage to aircraft occurring during the year with the maintenance costs involved per accident. Other accident/personnel injury data include: detailed injuries by types, accident types by work group, workers' compensation cost, occupational injury cost summary, disabling injuries (who-how-where), and passenger injuries.

The "AA Accident/Incident Report" is a complete index of all aircraft accidents, employee injuries and passenger injuries.[8] It also includes hours out-of-service to repair and repair (maintenance) costs. With this data kept on a yearly basis, AA's safety personnel can thoroughly review any trends that appear to be developing and put all the data to work in maintaining a strong and effective accident prevention program.[9]

Safety in the Maintenance Hangar and on the Ramp.[10] Constant attention must be directed to the potential hazards that exist in the hangar. General shop housekeeping, including floors and equipment, must be maintained in good condition. Floors should be kept clean and free of oil, grease, gasoline, water, or other hazardous and slippery materials. Both fixed and portable hangar equipment and machines, and electrical eiquipment in hangars and around aircraft, should be National Electrical Code (NEC) approved.[11] Housekeeping is an ever-existing problem in hangars. The nature of the work performed creates oil spills, debris, and cleaning materials on work surfaces. Oily rags, waste, and other materials saturated with flammable substances should be disposed of in metal containers equipped with self-closing lids. Designated aisles, walkways, and exits should be kept clear of all obstacles. Good housekeeping in hangars, nose docks, ramps, and other aircraft maintenance areas is essential to personnel safety and efficient aircraft maintenance.

Horseplay should be strictly forbidden, particularly around aircraft and during maintenance operations. Not only is "fooling around" dangerous to personnel, it can and has resulted in extensive damage to aircraft and in serious personnel injuries.

Personal injury hazards are present in several forms when aircraft are being cleaned. Falls can be prevented by not allowing personnel to climb on wet surfaces while the aircraft is being washed. Personnel should wear adequate protective clothing and chemical type cover goggles when cleaning aircraft, particularly when the cleaning involves the use of chemical brightening agents—caustics, acids, and phenolic compounds. Only authorized cleaning agents should be used.

When authorized flammable solvents are used in cleaning aircraft, all potential sources of fire should be prohibited within 50 feet of the operation and warning signs should be posted. The aircraft should also be effectively grounded at all times to prevent the collection of static charges. When aircraft are being washed or cleaned with flammable solvents and electrical storms are in the vicinity, the operation should be suspended immediately.

In maintaining aircraft, the following major sources of maintenance hazards should be considered.

Fires and Explosions. Both aircraft and flight line/ramp facilities present potential fire and explosion hazards. Gasoline, jet fuel, cleaning compounds, oxygen, and power equipment are typical of such hazards. In the interest of fire prevention, maintenance personnel should be familiar with the more common causes of petrochemical fires, especially those which involve flammable liquids. Before a fire can occur, three essentials must be present: (1) fuel in the vapor form; (2) air (oxygen); and (3) a source of ignition (Figure 33).

Because of the nature of the liquid products handled, fuel vapors may be present in the proper proportion with air to support combustion. Therefore, it is important to control all sources of ignition where such flammable or explosive vapor-air mixtures may be present. All employees involved in aircraft fuel servicing operations should be trained in fire-fighting methods and tactics and the use of appropriate fire extinguishers.

Falls. Maintenance, airline services, and aircrew personnel are exposed to the dangers of falling when servicing aircraft from workstands, platforms, and when working in or on the aircraft itself. The fall hazard is particularly acute during periods of high winds, inclement weather, darkness, and when aircraft cleaning and painting are being done.

Safety and Maintenance 131

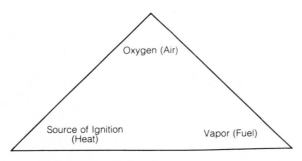

Figure 33. Fire triangle

Engine Operation and Repair. Engine repair personnel are often exposed to hazards when working on operating powerplants. Particular caution should be exercised when it is necessary to perform such maintenance on operating engines (see Figure 34).[12]

Electrical Shock. Electronics and communications equipment, power tools, and electrical supply lines and power units all present a definite danger.

Toxicity and Skin Contaminants. Many of the fuel and combustion compounds, as well as fire extinguishing agents, used in modern aircraft are harmful when inhaled or come in contact with the skin.

Aircraft Jacking and Tire Changing. Serious injury to personnel and extensive damage to aircraft can result from unsafe jacking operations. Tire changing can lead to strains and other injuries when performed under unsafe conditions.

Tools and Air Compressors. Both hand and power tools are sources of personnel injury when used incorrectly or without adequate safeguards. Air compressors are potentially hazardous when operated without regard for established safety standards.

Fire and explosions are ever-present hazards in aircraft maintenance and handling operations. New facilities and alterations or repairs to existing facilities should conform to the National Electric Code and the National Fire Codes. No source of ignition or any agent capable of igniting flammable vapors or gases should be permitted within a hazardous area as defined by the National Fire Codes.

Aircraft maintenance operations conducted within hangars, docks, paint facilities, or wash racks, ramps, and so forth should be accomplished in accordance with the National Fire Codes. Electrical devices and power equipment used in the hazardous areas should be approved explosion-proof types, including motors, switches, fixtures, and extension lights.

Smoking should be prohibited on the flight line, on ramps, in hangars, or on aircraft parking areas (including the inside of vehicles, and personnel

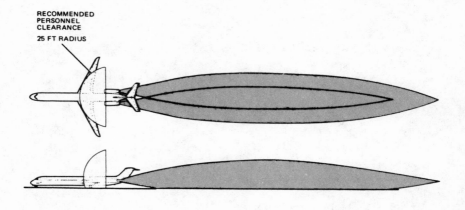

	DISTANCE BEHIND AIRPLANE			
	25 FT	50 FT	100 FT	250 FT
IDLE EXHAUST VELOCITY	55 MPH	45 MPH	30 MPH	----
BREAKAWAY EXHAUST VELOCITY	85	75	50	30
TAKEOFF EXHAUST VELOCITY	200+	200	150	75

EAR PROTECTION RECOMMENDED WITHIN 200 FEET OF RUNNING ENGINES TO AVOID HEARING DAMAGE.

FOR ADDITIONAL JET BLAST INFORMATION SEE DOUGLAS REPORT 67264, DC-9 AIRPLANE CHARACTERISTICS FOR AIRPORT PLANNING.

Figure 34. Restricted areas during engine operations, June 1982 (Courtesy of Douglas Aircraft Company)

should not strike matches or operate mechanical lighters, except in areas designated by the local fire marshal). Smoking should be permitted only in designated safe locations. No Smoking signs should be conspicuously posted where flammable liquid vapors are normally present.

Spilled aircraft fuel is a highly potential source of fire. Extreme care should be taken to avoid spilling fuel. If a spill or severe leak does occur around aircraft, all maintenance operations should be stopped immediately. Every effort should be taken to preclude spilled fuel entering storm or sanitary sewers as this can affect sewage treatment facilities and may result in a destructive explosion in the sewage system. If resulting fuel spills are washed down floor drains, sufficient water should be used to be certain that

Safety and Maintenance 133

all fuel has been flushed from the drainage system. Fuel spills are not only potential fire and explosion hazards but are also a waste of valuable energy assets. All fuel spills should be investigated to ascertain whether the spills were due to carelessness or equipment malfunction.

At no time should maintenance personnel clean aircraft parts, hangar floors, equipment, or clothing with gasoline or other unauthorized solvents. All cleaning should be done with approved high flash point or nonflammable substances or liquids. Flammable liquids, when stored in buildings, should be in approved containers, and stored in areas specifically approved by the local fire marshal.

Personnel maintenance stands and related equipment used on flight lines, ramps, and in hangars should be stored in hangars or effectively secured outside to prevent collision with aircraft, vehicles, or other equipment. Responsible supervisors should inspect daily the stands' brakes, jacks, wheel locks, securing cables and anchor connections to make certain the stands cannot accidentally be moved by wind, propeller wash, or jet blast.

Extensive damage to aircraft and serious injury to personnel has resulted from careless or improper jacking procedures. To ensure the safety of both aircraft and personnel, all persons involved in jacking operations should be thoroughly familiar with the applicable handbooks for the particular aircraft involved. Jacks should be operated, maintained, inspected, and tested in accordance with appropriate technical manuals. As an added safety measure, jacks should be inspected before use to verify lifting capacity, proper functioning of safety locks, condition of pins, and general serviceability. Before aircraft are raised on jacks, all work stands and other equipment not designated to be under the aircraft during jacking should be removed. If the normal weight and balance conditions of the basic aircraft have been disturbed by removal of heavy items, weight should be added where required to reestablish the balance before jacking the aircraft.

Aircraft may be supported on jacks outside hangars when wind conditions permit. When specifications for wind velocity for a particular aircraft are not available, maintenance personnel should accept a velocity of 15 miles per hour as the safe maximum for outside jacking operations.

When mounting or removing heavy aircraft tires, maintenance personnel should use tire dollies or other appropriate mechanical devices. Sufficient manpower is also necessary in addition to mechanical aids to handle safely heavy tires and wheels. Tire cage guards should be used during the inflation of tires. These guards will prevent injuries to maintenance personnel if destructive disintegration of wheel or tire takes place or if the locking ring breaks loose during inflation. Tire temperatures are more critical as aircraft become faster and heavier. Nitrogen gas, rather than compressed air, is often preferred for inflating tires, because the oxygen in air reacts with the tire rubber at high temperatures and pressures, resulting in deterioration and decreased tire life, and possible presenting a blowout hazard.

Fueling Responsibilities. All aviation fuels are classified as flammable-combustible liquids. There are two grades of gasoline and four grades of JP (jet propulsion) fuels commonly used as aviation fuel. In addition to the fire hazard, JP fuels may irritate the skin on contact and may cause dermatitis after repeated exposure. Fuels and maintenance chemicals are toxic! JP fuels are toxic if swallowed or inhaled.

Fueling aircraft from either trucks, hydrants, or pits requires the utmost caution on the part of all personnel. Modern fuels used in high performance aircraft are extremely flammable and easily ignited. Fuel vapors can be ignited by static or friction sparks, hot exhaust pipes, lighted cigarettes, electrical devices, and similar ignition sources. Jet fuels have an auto-ignition temperature of approximately 450 F, and aviation gas only slightly above 800 F. When gasoline vapor and air mixes between 1.5 and 7.0 percent by volume of gasoline in the air, ignition will result in explosion and fire. Fuel vapors being heavier than air will settle to the ground and accumulate in dangerous amounts in depressions, troughs, and pits. The danger area exists not only in the immediate vicinity of fueling operations but may extend downwind depending on weather conditions. The fuels used in modern jet aircraft are potentially more dangerous than gasoline and must be handled with the same respect and care as high-octane aviation gasoline.

Fueling Aircraft.[13] A written SOP for aircraft fuel from facilities on the particular airport shall be available in the airport manager's office and in each fueling truck. This SOP should include all the safety requirements of NFPA (National Fire Protection Association) Standard 407 and the particular petroleum company's standards. An abbreviated typical fueling procedure is outlined below.

A. Fueling Procedures—General

1. In order to service the aircraft promptly and efficiently upon arrival, the fueling supervisor should obtain as much advance data as possible.
    (i) Aircraft arrival and departure times.
    (ii) Estimated quantity and grades of product required (fuel, oil, etc.).
    (iii) Determine fueling method (hydrant, mobile refueler, over-wing, underwing, etc.).
2. Upon aircraft arrival, obtain quantity and grades of products required from airline or aircraft personnel.
3. The aircraft should not be approached until it is stationary, the main engines have been shut down, and it is ready for servicing.

Safety and Maintenance

4. The fueling vehicle should be positioned with a clear path to permit rapid removal in the event of an emergency and to facilitate egress upon completion. Consideration should be given to the location of the fueler's engine and location of the aircraft's fuel vent system. The fueling vehicle should not be positioned where it would obstruct aircraft exits and loading areas.
5. All fueling operations should be conducted outside hangars or similar enclosed buildings.
6. After the fueling vehicle is in position, fire extinguishers should be readily available in accordance with NFPA standards.
7. Grounding/bonding.
    (i) Attach a grounding cable from the fueling vehicle to a satisfactory ground connect.
    (ii) Connect a grounding cable from the ground to the aircraft fitting, if one is provided, or any convenient unpainted metal point on the aircraft.
    (iii) Bond the vehicle to the aircraft. Where a "Y" or "V" cable permanently attached to the fueling vehicle is used to accomplish (i) and (ii) above, a separate bonding cable is not necessary.
    (iv) Bonding and grounding requirements and electrical continuity checks shall be in accordance with applicable NFPA standards.
8. Hoses should be run out on selected routes which will prevent them from being run over by serving vehicles. Kinking and twisting of hoses should be avoided. Pressure fueling couplings and overwing nozzles should not be dragged over airport surfaces. Dust caps shall be in place at all times when coupling and nozzle are not in use.
9. Before fueling, check with the airline representative to confirm that all pertinent equipment on the aircraft is positioned ready to receive fuel.
10. If fueling is performed at night, the fueling area should be well illuminated.
11. Take fuel sample, if required, under full flow conditions. Ensure that the sample is representative. Also, when required, perform water detection test on sample.
12. If an aircraft is fueled with passengers on board, an airline representative should be on board to ensure that No Smoking rules are observed.
13. No fueling shall be conducted during any aircraft maintenance that might provide a source of ignition to fuel vapors. All radio and radar equipment must be "off" and switches must not be manipulated.
14. While under full flow, check vehicle and fuel system for leaks, etc.;

observe that the filter differential pressure does not exceed acceptable limits. If leaks or signs of leaks occur, all fueling operations must be halted immediately.
15. Flow rate for fuels with flash points below 100 F (37.8 C) (for example JP-4) should be reduced from the normal rates to decrease static electrical buildup. (See NFPA Standard 407 and FAA Order 8110.34A.)
16. The operator should be positioned at a point where there is a clear view of the vehicle control panel and aircraft fueling points. The deadman control must always be used and should never be wedged for blocked open or positioned to defeat its purpose.
17. No Smoking signs must be displayed in prominent positions near the aircraft and fueling vehicle.
18. Unauthorized persons are not permitted in the fueling area under any circumstances.
19. Fueling an aircraft which has one engine running in a nonroutine emergency operation. Because of its nonroutine nature, the operation must be reviewed beforehand by the airline and fueling company representatives. Fueling should only be performed if the operation is within the scope of the current airport regulations and all prescribed safety precautions are followed. Fuel must only be loaded on the side opposite to that of the running engine.

B. FUELING PROCEDURES—UNDERWING SERVICING FROM FUELERS AND HYDRANT VEHICLES

1. Ground and bond fueling vehicle to ground and to aircraft.
2. Hydrant servicers only:
   (i) Open hydrant pit cover (check the product grade before connection).
   (ii) Place Warning signs or lights in position at hydrant box.
   (iii) Remove dust caps from valve in hydrant box and from coupler of inlet hose.
3. Open aircraft fueling station access door and remove dust covers from hose nozzle and aircraft valves.
4. Connect delivery hose nozzle to aircraft fueling point, open nozzle and place appropriate aircraft fuel switch to the "on" position, connect the hose coupler to the hydrant valve after checking both valve surfaces to be sure they are clean and dry, then open hydrant coupler and adapter and activate fueling vehicle with deadman control.
5. Start fueling—keep alert and take all precautions for safety and be sure not to exceed aircraft structural fuel pumping pressure.

Safety and Maintenance

(i) Continually monitor the underwing fuel gauges and be in a position to shut off quickly flow in an emergency.
(ii) Never block a deadman valve in the "on" or open position. Under no circumstances shall the nozzle be left unattended during fueling.
6. Never overlook the possibility of an accidental fuel spill or leak from the aircraft or the fueling vehicle.
7. Upon completion of fueling:
   (i) Check fuel quantity dispensed with fuel quantity requested.
   (ii) Disconnect hydrant coupler and stow hoses.
   (iii) Disconnect hose nozzle and replace dust caps.
   (iv) Close fueling station access door.
   (v) Remove ladders or lower platform.
   (vi) Remove bond cable from aircraft to fueling vehicle and ground cable from ground to fueling vehicle.
8. Check the filter/separator sump on the fueling vehicle for water following the fueling. If an unsatisfactory check is found, request an airline representative to check aircraft sumps, drain any water found, and acknowledge that aircraft is water free.
9. Remove fueling vehicle from aircraft area as soon as possible after servicing is completed.

C. Fueling Procedures—Overwing

In addition to the procedures given for underwing, where applicable, the following should also be applied.

1. Always use suitable ladders and mats to avoid damage to aircraft wing. Use extreme care to prevent hose or nozzle from damaging deicer boot or leading edge of wing.
2. Set wing mat in place.
3. Connect static bonding wire from nozzle to receptacle, post, or other metal part of plane before opening fuel tank cover.
4. Open tank access, remove nozzle dust cap, and insert nozzle, keeping constant contact between the nozzle and the filler neck while fueling.
5. Start fueling—overwing nozzles should not be equipped with "hold open ratchets," which lack will prevent such nozzles from being unattended during delivery. Make frequent visual checks of tank capacity, taking extreme care to prevent spillage or overfilling.
6. Upon completion of delivery, quantity in tank should be checked with fuel quantity requested.
7. Replace and secure tank access caps. Disconnect nozzle static bond wire. Replace nozzle dust cap.
8. Return hose to fueler reel.

All airport equipment should be marked to identify the type and grade of aviation fuel being issued and dispensed in order to preclude intermixing or contaminating the fuels. One of the most common accidents related to aviation fuel is the intermixing of jet fuel in an aircraft that requires aviation gasoline (AVGAS). A reciprocating engine will not operate on jet fuel.

A. Fueling Systems. Airport fueling systems should be marked utilizing the marking code described in the subsequent paragraph. Particular attention should be given to marking pumps, valves, and the lines used for loading and unloading fuel. Where space will not permit banding and printing the name on the pipe, fuel service hydrants, hydrant carts, hydrant covers, and pits which hold valves, hydrant connections, hose reels, filters, and other fueling equipment should be painted in the identifying product grade color. Piping systems that are buried or inaccessible should have all exposed valve stems and wheels painted the identifying product grade color and a flag post permanently fixed to the pipeline or a concrete pad near the valves showing the marking code.

B. Fueling Vehicles. To prevent error in identification of fuels in fueling vehicles, marking as shown in Appendix 3 of AC 150/5230-4, with white letters at least three inches high, should be painted at the hose outlets and on the doors of the vehicles.

The National Transportation Safety Board has investigated recent accidents involving Piper PA-31-350 airplanes which required aviation gasoline (AVGAS) but which had been inadvertently serviced with jet fuel. One accident occurred on April 18, 1982, at San Antonio, Texas; and another occurred on May 22, 1982, at Midland Texas. These accidents occurred shortly after takeoff, and the airplanes crashed in densely populated areas (listed in an NTSB letter to the General Aviation Manufacturers Association [GAMA], dated October 20, 1982).

During the past three years, there have been 10 reported aircraft accidents caused by improper fuel. Many times, contaminated fuel has been found during preflight inspection, and the fuel tanks drained prior to engine start. The aircraft involved in the accidents included light single-engine personal-pleasure-type aircraft, agricultural aircraft, and twin-engine executive-type aircraft. All aircraft are susceptible to being serviced with wrong fuel. In addition to loss of life and injuries, all of these type accidents, incidents, and errors are costly, embarrassing, and preventable.

A marking code should be used to permit rapid identification under varying visibility conditions. The code is comprised of three systems described below.

A. Fuel Naming System. (1) Aviation Gasoline. The naming system for the four grades of aviation gasoline is made from the general term "AVGAS" followed by the grade marking. The grades are identified by their performance numbers as recognized by all military and commercial specifications; i.e., 80, 100 LL, 100, and 115. AVGAS is a widely used abbreviation of the words "aviation gasoline." The use of the naming system "AVGAS 100" indicates that the aviation gasoline within an airport fueling system meets the minimum requirements of the United States military or NATO specifications for that grade.

(2) Jet Fuel. The three classifications of aviation turbine fuels are nearly universally referred to as "Jet Fuels" and are generally described as JET A, JET A-1, and JET B. They are used in "turbojet" and "turboprop" engines. These three classifications are: *Jet A*—a relatively high flash point distillate of the kerosene type, having a −40 F freezing point (max); *JET A-1*—a kerosene type similar to JET A but incorporating special low temperature characteristics for certain operations, i.e., −53 F (−47 C) freezing point (max); and *JET B*—a relatively wide boiling range volatile distillate having a −58 F (−50 C) freezing point (max).

In some cases, it may be desirable or necessary to indicate on the marking system some additional refinement. Words can be added below the "JET A" to cover these requirements. Also, some manufacturers may desire to show their product brand name on the airport fueling system for a transition period. The brand name could follow behind the product type identification and be separated.

B. Color Code System. (1) Aviation Gasoline. The naming system for aviation gasoline grades is printed in white letters and numbers on a red background. Red was chosen for the background because it is an indication of the special care which must be taken in the handling of the more volatile fuels.

(2) Jet Fuel. Jet fuels are of two distinct types. The JET A and JET A-1 are low-volatility fuels of the kerosene type. JET B is a widecut gasoline type made from parts otherwise used to make both kerosene and gasoline. These three grades vary in color from white to light yellow. Jet engines will operate on all three grades under most conditions, but each grade of fuel has characteristics which require that it be left separate. It is also desirable to make a clear distinction between the low volatility JET A and JET A-1 fuels and the more volatile JET B. The naming system for the jet fuels is painted in white letters on a black background in contrast to the gasoline color code which is painted in white letters on a red background.

(3) Banding System. Circular bands of an identifying color are painted or taped around the piping at intervals as one part of the marking code. They are for use both adjacent to the naming system and by themselves. The circular band was chosen because it appears the same from all directions. *(1) Aviation Gasoline.* The color of the single band around the piping or hose is the same color as the dye in the grade of AVGAS flowing through the line. They are red for AVGAS 80, blue for AVGAS 100 LL, green for AVGAS 100 and

purple for AVGAS 115. A minimum four-inch-wide band is recommended. If the pipeline is painted the color of the AVGAS, then no banding is needed. *(2) Jet Fuel.* Black, gray, and yellow bands are used to identify JET A, JET A-1, and JET B, respectively.

During any fueling operation, the area within a radius of 500 feet around such an operation should be considered a restricted area. Personnel engaged in the actual fuel servicing operation should be instructed that clothing can generate and accumulate static electricity (for example, wearing a wool sweater under a nylon jacket or a combination of wool and a synthetic fabric). In this regard, cotton garments have the least potential for generating a static charge.

Aircraft defueling should, at all times, be accomplished under controlled supervision and conditions similar to fueling operations. During defueling, aircraft should be no less than 50 feet from any building, smoking area, source of ignition, or other aircraft, and so situated that fuel vapors cannot be wind carried to any ignition source (the fuel-servicing vehicle being the only exception). The aircraft should be statically grounded as specified by FAA requirements or aircraft manuals.

Defueling should not be permitted when electrical storms are in the immediate vicinity. Prior to defueling, personnel should remove the fuel cell filler cap and make certain the vent line is open. To ensure absolute safety, no other work should be done on the aircraft during defueling. The inerting process should be started after defueling and completed before the tank access door is opened. Maintenance personnel should exercise extreme caution during inerting, making certain the N_2 (nitrogen) hose line is effectively grounded. The depuddling operation is extremely hazardous because it requires maintenance personnel to enter cells to remove fuel residue. For maximum safety, personnel entering a tank cell should work at least in pairs, one of whom should remain outside as a safety guard. The safety guard should never enter the fuel cell in a rescue attempt until additional help is present and then only when equipped with an air-supplied respirator.

Maintenance personnel should wear white cotton duck overalls that have no metal accessories or buttons. After fuel residues have been removed from the tank cells through depuddling, maintenance personnel should airpurge the cells. The cells' atmosphere during purging should be checked periodically with an approved and properly calibrated testing instrument. Air blowers should be operating at all times when personnel are working on fuel cells.

Short circuits, overloading, accidental grounding, lack of equipment grounds, poor electrical contacts, and misuse are all responsible for the major accidents involving electricity. Electrical control boards, switches, transformers, and other hazardous electrical devices should be located where the

Safety and Maintenance

possibility of accidental contact is minimized. Grounded railing, barriers, or enclosures should be used to protect conductors, bus boards, switches, control panels, and other hazardous electrical facilities, particularly around high voltage fixtures and transmission lines. Electrical cords should preferably be heavily insulated. Personnel should avoid excess bending, stretching, and kinking of electrical supply cords. Extension cords used with portable tools should be of the three-wire type with three-prong plugs except when using double insulated tools (see Figure 35 for typical aircraft electrical ground service connections).[14]

Personnel working around electrical circuits, regardless of location (flight line, ramp, shops, and so forth) should not wear rings, watches, metal-rimmed glasses or other metallic objects that could act as conductors of electricity and cause shock or electrocution. To provide immediate help to persons who have touched live circuits, it is essential that all electrical workers be trained in the approved method of mouth-to-mouth resuscitation, which should be started at once and continued until the victim has begun breathing again and medical aid has arrived on the scene. If the heart stops, closed-chest heart massage must be started immediately.

With the rapid advances in laser technology and increased use, the probability of personnel exposure to injurious intensities of laser radiation is greatly increased. The hazards that exist from operation of lasers result from the concentration of high energy upon body tissue. The effects on personnel usually consist of eye or skin damage. In addition, physical hazards exist, such as exploding flash lamps, wires, and chemicals. Also, X-rays and rapid frequency radiation hazards may be associated with the operation of lasers. Safety precautions in laser operations include: (1) Never look directly into the primary laser beam. The high intensity of most laser beams can cause irreparable damage to the eye. (2) Avoid looking at laser beam reflections. (3) Do not place any part of the body in the laser beam path. Severe burns can result in spite of protective clothing. (4) Wear the safety goggles that are designed for the frequency or wave length of the laser being used.

Microwave and radar radiation, when of sufficient intensity, can damage human tissue (particularly the eyes). Radar and microwave equipment is a common source of radiation. Radiation is also dangerous near fueling operations and other flammable materials. Within recent years, the use of microwave ovens in kitchens, cafeterias, and certain models of aircraft has become widespread. Inspection of these ovens has, in some cases, revealed leaks of dangerously high intensity rays.

Because of the high vacuum present in cathode ray tubes, operators and technicians should be particularly careful not to bump, crack, or scratch this type of electronic component. There is a great danger of implosion when a cathode ray tube is damaged or handled too roughly. Large vacuum tubes should be stored in cartons or cases to protect them against breakage and personnel against injury from flying glass if implosions occur. In addition to the hazard of serious lacerations from flying glass, personnel may also be exposed to radioactive materials which may be assimilated, ingested, in-

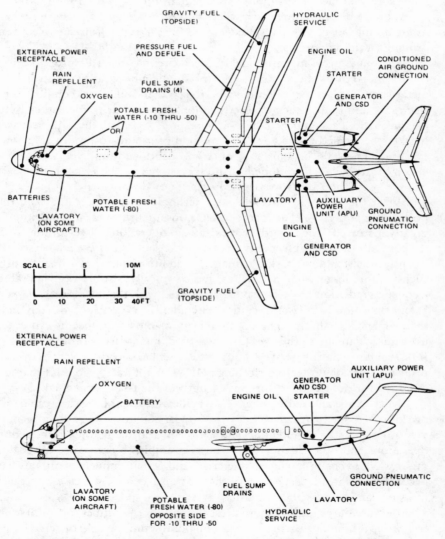

Figure 35. Typical aircraft ground service connections, June 1982 (Courtesy of Douglas Aircraft Company)

haled, or absorbed, thereby presenting a serious internal radiation hazard. Such radioactive materials may also be deposited on the surface of the skin and produce contamination.

Electrical circuits should be grounded at the exact point of intended repair or adjustment before the technician actually makes contact with that particular area of the equipment. Grounding or shorting sticks should be used and should not be removed until repairs are completed. Also, wooden, fiberglass, or other nonconducting polymer plastic stools and ladders should be used when working on electrical or electronic equipment.

Compressed air is used in many shop operations for spray painting, tire inflation, fuel line cleaning, etc. Many mechanics are tempted to use air under high pressure to clean work benches, vehicle chassis, and other parts. Air cleaning is injurious to the eyes. Mechanics should never engage in horseplay with compressed air. Do not exceed 25 psi for cleaning purposes, and chip guarding should be provided.[15]

Spray booths should be used in all paint shop operations to localize fire and explosion hazards. Forced-air ventilation should be provided in all paint spray booths to prevent the accumulation of flammable and injurious vapors in the atmosphere. All sources of ignition should be removed from the paint shop. Electrical equipment and fixtures should be explosion-proof and effectively grounded at all times. Smoking should not be permitted in the paint shop. Aircraft must be grounded properly before painting or paint removing. All metal- and fabric-covered objects that could produce static charges should be grounded effectively or bonded before spray painting is authorized. Suitable fire extinguishers should be provided at all painting and paint-removing locations. The NFPA (National Fire Protection Association) normally requires that overhead sprinkler systems be installed in permanent paint shops.

The battery shop facility should conform to the provisions of Articles 480 and 500, National Electrical Code. Operating personnel are exposed to the possibility of painful chemical burns, explosive gases, and toxic chemicals. Battery shop workers should be thoroughly trained in their jobs and indoctrinated in safe operating procedures. Operating procedures should be posted. Nickel-cadmium and silver-zinc batteries should be serviced in an area isolated from lead-acid batteries. Sufficient ventilation must be provided to prevent acid fumes from entering the nickel-cadmium or silver-zinc battery shops.

Lead-acid battery charging equipment should be located in properly ventilated rooms. Excessive charging of lead-acid batteries should not be permitted as hydrogen gas is generated.

When handling electrolyte, shop personnel should wear protective chemical goggles or full-face shields, rubber gloves and aprons, and acid resistant safety shoes, or rubber knee length (safety cap) boots. When mixing, always pour electrolyte into water. Never pour water into electrolyte. The heat of dilution will cause water to boil and spatter when poured into electrolyte. Running water should be available immediately during diluting operations. Treat acid burns with baking soda and water. Treat hydroxide burns (alkaline batteries) with vinegar and water.

In larger battery shop operations, the room/area should be well ventilated to prevent accumulation of explosive gases or toxic vapors. Floors should be resistant to or protected from electrolyte accumulation. Material/equipment should be provided for neutralizing or flushing spilled electrolyte. Facilities for quick drenching of the eyes and body should be provided. No Smoking signs should be posted in the area. Fire extinguishers should be provided. Electrolyte should be mixed in a well-ventilated area. Acid or alkaline should be poured gradually, while stirring, into the water. Water should never be poured into acid solutions.

Unsafe conditions of installations, buildings, and grounds are hazardous to worker personnel and to visitors. A preventive maintenance program should be designed to meet the objective of safety. Building doors should open outward. Exits should be clearly marked and lighted. Illuminated exit signs should be provided and installed for emergency exits and passageways as required by NFPA and local building codes. Doors should not open immediately on a flight of stairs but on a landing at least the width of the door.[16]

This chapter has briefly reviewed the scope of safety in aircraft operations and maintenance. It has been shown that the FAA's primary mission is directed toward aviation safety in the air and on the ground. The FAA safety objective starts when a manufacturer first initiates a new aircraft design and doesn't stop until an aircraft in the system is grounded—permanently. The NTSB was acknowledged for the mission it performs, that of serving as the overseer of all United States transportation safety. Its specific mission is to improve transportation safety and its safety recommendations to agencies such as the FAA carry considerable weight. The chapter further reviewed a major airline safety program that is total in its quest to provide maximum safety to its flying passengers and flight crews, to its ground maintenance crews and to its many employees working in other capacities within the airline. Lastly, a detailed review of safety practices for the maintenance hanger and ramp was included. Electrical repairs, refueling, defueling, and electronic equipment maintenance are high among the work areas where individuals must adhere rigidly to strict safety rules.

11

Electronic Data Processing

ELECTRONIC DATA PROCESSING (EDP) is becoming more useful and prominent in the aviation industry as the input of complex data increases, the requirement for customer satisfaction continues to be important, and the fact that high-speed computers are starting to sell at a price that any company, large or small, can afford. In addition to its advantage in output retrieval speed, the computer has an unfailing memory and an accuracy far beyond human capacity.

The data-processing system can be set up so that data can be programmed to any aspect of maintenance or other aviation information that is desired. The system can effectively and efficiently be used for: labor—schedules for different types of maintenance requirements, performance—standardizing man-hours for each job and determining the labor effectiveness, equipment—machine processing to include lubrication requirements, downtime, and total equipment/machine records, and costs—cost control data on all aspects of maintenance.

Since many industries have turned to the computer, it is not surprising that specialty companies have come into the maintenance field with some good ideas for electronically working the data in maintenance categories. Two such companies, Aviation Information Services, Inc., and CAMP Systems, Inc., have jointly developed a comprehensive aircraft maintenance program (CAMP/DISPATCH) aimed at the commuter airline market that automates all aircraft maintenance records so that information on scheduled and unscheduled maintenance collected internally by the maintenance manager can be entered directly into the flight department's own computer system. Designed to meet FAA requirements for Part 91 or Part 121, 123, or 135 operators, the features of the CAMP/DISPATCH maintenance system include:

The accommodation of a separate maintenace plan for each aircraft. The system keeps track of the prescribed plan for each aircraft in the fleet, covering FAA requirements and manufacturers' recommendations.

Maintains a maintenance history for each aircraft (Figure 36). Full maintenance records on each aircraft are permanently stored in the system, and can be retrieved instantly. Information is also stored by type of maintenance so questions like, "When was the last pressurization rate switch inspection?" are answered immediately (Figure 37).

Flags time-saving options. If, for example, unscheduled repair on a landing gear is required, the system can flag related maintenance operations that are scheduled in the near future. So the option assures that all the work is done at the same time, instead of bringing the same plane back for scheduled maintenance at some later date.

Provides an optional inventory control system for parts. The system keeps track of all parts in inventory and lets the manager know when it's time to reorder. It issues an "overdue" report when parts are not received from the manufacturer, or overhaul activity is not accomplished by a specified time. It maintains a history of individual components—when they were installed, when they were removed, and why. The system includes stockroom transactions, repair history, and monitoring of items with limited shelf life.

Another company taking aim at the commuter airlines is a new software producer, Airline Software, Inc. (ASI), which is designing easy-to-use systems for companies short on in-house programming expertise and manpower. Designed for use with the IBM System 34, which ASI's founder estimates is operated by 80 percent of all United States commuters, the software systems are what is called "user friendly." They deal in plain, job-related language "instead of computer language." "That means that the user need not know anything about data processing to use the system successfully," he claims.

Scheduling, inventory, accounting, crew information, customer lists, flight following, inspection maintenance, and revenue accounting are all systems currently covered by computer software. All may be used alone or interactively. And all are "exception based." The user gets a choice of summary or detail, not a full status readout every time a request is made. The flight-following system, controlled by the scheduling system, also produces

Electronic Data Processing 147

Figure 36. Aircraft maintenance history: abbreviated list (Courtesy of Aviation Information Services, Inc.)

Figure 37. Aircraft status report (Courtesy of Aviation Information Services, Inc.)

hard copy as well as a CRT display of each day's schedule, aircraft assignments, and status. This enables dispatchers to monitor from single sheets instead of from the customary wall-to-wall status board. The revenue accounting system controls much besides accounting: interline clearing-house activities, ticketing and rate information, and other revenue-related functions.

One of the most significant and profound developments of the technological revolution is the electronic digital computer. Computers are used in a wide range of fields including the agricultural and the medical and in all the steps of the industrial process. With the help of man, they can take diverse and large quantities of information and transform the material into usable forms. The computer itself cannot physically do any maintenance. However, it is an important part of maintenance because it is the most modern device designed to facilitate information processing in industry. The computer can, through an electrical signal, close a valve if a function down the line malfunctions, but it cannot change a part that might need to be changed. The ways the computer gets its information and how important information is received from the computer will be discussed.

There are several ways in which the computer gets its information. The off-line method is where the computer receives its information manually from a human operator. The data used this way can be sent from instruments punched onto cards and then fed into the computer. This way is usually the slowest and is often subject to human error. The in-line mode of processing information is quicker than that of the off-line mode because the operator enters process data rapidly and frequently directly into the computer through a nearby keyboard. In both of these types of methods, the output information is usually presented as a printed message or another type of visual display.

An automatic method of collecting information is called the on-line mode. In this type of system, the system is connected physically to the process in order to receive information without human intervention or delay. This is normally done by sensors in the system that can detect malfunctions. The on-line system can be broken down into two categories, closed loop and open loop. In the closed loop, there is no manual intervention in the operating mode and the control actions calculated and recommended by the computer are directly applied to the process. In the open loop system, an operator is interposed between the computer and the process to interpret the computer output and apply control decisions to the control of the process.

The question, of course, is what good is the information if one does not know how to use it. The way it is applied to maintenance involves the processing of the paper work into the computer. The information becomes invaluable to the maintenance personnel. For example, this information can tell if a piece of equipment is malfunctioning more often than it should. This could lead to a design check and may show that a manufacturing deficiency exists. Also, the information can be used to schedule maintenance on an "as needed" basis. It can keep records of what has been done and when it was accomplished. When it is used as a maintenance scheduler, it aids in the efficient work flow of a shop. The recording of maintenance that has been done can be used as a quality assurance tool to verify what malfunctions keep

Electronic Data Processing

coming up. It can also affect the maintenance shop work load by indicating how many people are needed to complete any given work load. In certain industries, computers monitor the workings of a system and indicate when certain maintenance needs to be done.

Even though the computer may seem like a miracle breakthrough from a technical standpoint, it must be judged by the same standards as any other project which competes for limited capital funds and personnel. There are four major factors that contribute to the cost of a computer. First is the cost of the computer itself, then, two, there is the cost of any instrumentation that may be necessary, third, the cost of installation and, fourth, there is the cost of staffing for the computer. Of all the costs incurred in the installation of a new computer system all but one is a continuous cost. The cost of operating the computer will always be present, but all the other costs incurred are definitely paid back in savings in man-hours and equipment. Those savings are continuous and can be improved upon as new and better systems are invented. An example of the way a computer can save money is by the immediate availability of results that may be needed, which can assist in greater accuracy and more rapid decisions.

There are a number of intangible savings that may be cited also. Better plant management and supervision is possible under computer control. By reducing the effort needed in logging data and maintaining routine paperwork, the computer reduces the quantity of personnel man-hours required. The computer frees the supervisor of these tasks so that he or she will have more time to run his or her shop more efficiently. A short-range benefit of computer operation is increased knowledge that can be applied through the organization. The computer can help management develop new processes for doing things. These are just a few of the invaluable ways a computer can be used as a cost-cutter and money-saver.

The more uses for which the computer is used, the lower the unit cost or the total cost that will be attributed to the maintenance department. On the other hand, the company statistician can show where the EDP equipment has actually saved the company in terms of labor, time, and money. The current types of computers are so varied that almost every activity in the aviation company can be computerized efficiently.

One example of an EDP-based aviation maintenance information system is the CAMP/DISPATCH (computerized aviation maintenance program).[1] The operation of the CAMP program centers around the exchange of maintenance information on a monthly basis. This information is exchanged in two basic formats common to all data retrieval systems, *Input* and *Output*. *Input*—Compliance data accumulated through the everyday operation of maintenance tasks is documented on the Maintenance Requirement Card formated to handle both scheduled and unscheduled maintenance. *Output*—This is a series of monthly updated reports pertaining to the current maintenance status, past maintenance history, and upcoming maintenance due, plus accompanying reports pertaining to fleet utilization and reliability updated at selected periods.

In developing a total maintenance system, CAMP Systems, Inc., has attempted to keep in mind five essential ingredients which make the program meaningful in regard to the reliability data available from it. The program must be well planned, allow for maintenance scheduling, provide a means of recording compliance data, ensure there are adequate quality control checks, and require only a minimum amount of file maintenance. The following approach has been taken for all aircraft models:

Planning—Each maintenance requirement has been identified according to ATA (Air Transport Association of America) 100 specifications and itemized in the form of a computer record. Inspection, servicing, and overhaul/replacement frequencies are in accordance with the recommended or mandatory requirements contained in the applicable airframe or engine manufacturer's Inspection Schedule (ATA 100, Chapter 5) or maintenance manuals. Service Bulletins, Airworthiness Directives, and Type Data Certificate requirements are researched and identified in a similar manner.

Scheduling—Each month the computer reviews the status of the aircraft and makes a comparison of frequency, compliance, and projected utilization in order to identify the individual tasks that will become due in the next sixty- or ninety-day period. These tasks are listed, according to urgency, on the Maintenance Due List and sent to each operator with the necessary Maintenance Requirement Cards to complete the required tasks. The scheduled maintenance is accomplished based on the aircraft's availability and a carbonless copy of the completed cards is returned to CAMP Systems, Inc., for processing—which completes the loop. Each operator maintains a sufficient supply of Maintenance Requirement Cards to ensure system continuity in the event of unscheduled maintenance.

Recording—Each maintenance task on the aircraft appears on a Maintenance Requirement Card along with the procedural steps necessary to accomplish the task extracted from the applicable technical manuals. Since the card contains scheduling information along with spaces to record compliance data and signature, it serves as a work order in addition to a permanent maintenance record. Only a carbonless copy is returned for processing with the original being retained by the operator as his or her permanent maintenance record. In addition, the sign-off area for recording compliance data has been structured to handle any combination of component replacement, inspection, or service in such a manner as to ensure the reliability of the overall status in regard to associated maintenance, repetitive Service Bulletins, or Airworthiness Directives.

Electronic Data Processing

Quality Control—Each piece of input data (maintenance compliance information) is screened for obvious errors or omissions by experienced analysts familiar with the maintenance of that aircraft model. The data is then key punched and key verified for computer processing. Computer software checks are prevalent to prevent invalid, insufficient, inconsistent, or otherwise incorrect information from getting into the reports making the resultant reliability information questionable. Experienced technical writers constantly review revisions to manufacturers' inspection schedules and maintenance manuals to ensure the most current maintenance procedures are maintained on the Maintenance Requirement Cards as well as apprising the operator of manufacturer changes in maintenance frequencies.

File Maintenance—File maintenance is kept to a minimum as each report (Status, Due, and History) supersedes the previous month's which can be discarded. FAA-approved permanent maintenance records are immediately available upon completion of each task as only carbonless copies are returned for processing, thereby eliminating the need for duplicate log book entries.

The preceding paragraphs touch briefly on only a few of the basic features of the CAMP program as it exists today. Since its inception in 1968, the program has undergone numerous changes and revisions in order to keep up with the improvements in aviation maintenance technology. Whatever course aviation maintenance takes in the years to come, CAMP Systems, Inc., has made a commitment to its customers to maintain the standard of a total maintenance system.

12

Aviation Maintenance Management Problem Areas

THE MAINTENANCE MANAGER IS CONTINUOUSLY confronted with a myriad of problems that involve aspects of technology, administration, production, human behavior, and include all of the management functions outlined in Chapter 5. Specific areas that concern him or her are: effectiveness (economic) of the equipment, facilities, and personnel; the working environment; OSHA safety requirements; manufacturers' requirements; FAA regulations and FAA inspectors; control of equipment, tools, and aircraft parts; inventory of maintenance supplies and parts; maintenance quality assurance; personnel training; inspections; subcontracting of maintenance work; paper-work administration and library requirements; pricing determinations; electronic data processing; personnel salaries, merit increases, etc.; theft or misappropriation; damage through mishandling; and controlling costs and profitability.

All businesses, large or small, service oriented or product oriented, have problems that management must deal with. In fact, that is what a manager is paid to do. His or her job is to keep things running smoothly on a day to day basis. If a problem arises, that manager must solve it as quickly and efficiently as possible and maintain the business as an ongoing concern. This section will focus primarily on some of the problems faced by maintenance managers. Not all of the areas discussed will be peculiar to the aviation industry; some

Maintenance Management Problem Areas

are common to all businesses. Whether or not the problem is a universal management problem is not the issue. The solution to the problem is the main concern.

This text cannot possibly cover all of the previously listed management concerns that deal with the maintenance activity. Therefore, selected problem areas were chosen for review. Presented here are only a few of the management "headaches" and what can be done to cope with them.

Consider first an actual situation in an aircraft maintenance department where the manager had control of the parts department as well. This situation will point out several of the manager's problems.

The operation of the parts department is one of the most rigidly controlled areas within the company. Aircraft parts are expensive, sometimes easily damaged by mishandling, and—in the case of instruments, radios, etc., that are common to all aircraft—constantly subject to theft or misappropriation. Establishing the present control was done in textbook fashion. First on the agenda was the inventory and labeling of parts. Second was verifying the filing system to locate and number the quantities on hand. One recurring problem is having one part that will fit several machines. For example, one generator that would fit six aircraft. The actual inventory was one, but the filing system reflected six. Needless to say, the difference between $200 and $1,200 on one item could drastically alter the true value of an inventory if the trend applied to other high value items. The really tough decisions were made in deciding the quantities to maintain. Common items like spark plugs, tires, instruments, hydraulic lines, etc., had an established use rate that was easy to verify through work orders, purchase orders, and vendor recommendations. The seldom needed, but critical items that often grounded the aircraft were difficult to identify. An elaborate system has developed within the industry to alleviate problems of this type. AOG (aircraft on the ground) purchase orders get top priority at all wholesale houses and factory depots. A few basic operating rules are: 1) do not stock over six month's supply of any kind, 2) customer special orders require a 20 percent deposit (many businessmen get burned in this area because of the difficulty in returning stock), and 3) high value (over $500) requires approval of the president.

One of a manager's main concerns in managing a business is to keep it operating at a profit. But even though the name of the game may be profit, some management practices cost money in lost profits and cash losses. The area of parts sales is a good example of this. If an average maintenance shop foreman were asked how to make a profit from selling parts, his or her reply would probably be to buy them at the lowest cost and sell them at the highest prices. Most shops take their direct cost and subtract it from their selling price, calling what is left over profit. Too bad it is not that simple.

There is an item, called "cost of sales" that is often overlooked by managers. "Cost of sales" takes into consideration equipment depreciation, return on investment, and the cost of borrowed money when figuring out profit margins.[1] If a maintenance manager does not know the profit percentage that he or she makes on parts sales, then he or she is not aware of whether he or she is making or losing money. Another way in which shops miss out on parts sales is that they fail to sell them. If a mechanic does not install a new set of plug washers when cleaning and gapping a set of spark plugs, which by the way, all manufacturers recommend, that is a missed sale of $1.50 to $2.00. A small sale but the totals can add up.

Here is another instance of what can happen if maintenance personnel are not parts sale oriented. This is an actual case.

> A service manager for a parts distributor gets a trouble call from an FBO. The operator chews him out because his company is selling bad vacuum pumps. It seems the FBO installed a new pump on a customer's airplane and it failed after only five hours. The pump was changed again and the same thing happened. The FBO went through four pumps before someone got around to calling the service manager. The problem was solved when the FBO mechanic answered the service manager's first question, "Did you change the filters?" His reply was no. Had the shop foreman or mechanic sold the customer a new filter for $25, not only would the FBO have made a little more profit, but they would have avoided the $2,000 loss that they took on this one job. It is not a good idea to sell a customer a lot of parts he does not need, but at the same time, you don't want to short change yourself either. It is always a good idea to review the parts and consumables that will be required to complete a job.[2]

Consumables and hardware are areas that can be looked into. This is an area where repair stations, too often, miss out on profits. Consider the lubricants, solvents, safety wire, and other items that get used in everyday maintenance. Some shops absorb these into their overhead but the best way to handle this is to allow a fixed amount for each job. Adding just enough to cover actual costs can turn a loss into a break-even job or make a small profit.

The way in which hardware is ordered can also provide an opportunity to make a profit. Instead of ordering small parts in small lots, order a large supply at one time. This saves time and expense involved in making out many small orders, and also takes into consideration that most suppliers' price schedules for hardware are based on quantity. For instance, a certain size nut may come in minimum quantities of 25, priced at $27 per 100. When ordered 1,000 at a time, the price may drop to $20 per 100. This is how quantity purchasing can increase the profit margin on small parts.[3]

> Another element of the parts ordering process is making sure that the correct part is ordered from the supplier. When the wrong part is ordered, the service facility not only has an angry customer on their hands, they also

end up wasting valuable time reordering the part. That is, if the customer continues to do business with them. Two mistakes usually seem to cause the most trouble. One is failing to give either a complete description of the part or a complete description on the aircraft it goes on. The other is ordering from an obsolete parts catalog.

Again, it seems that common sense is the prevailing factor in making FBO's maintenance facility a little more profitable. All a maintenance manager needs to do is raise his workers "parts consciousness" a little bit. But this is usually easier said than done. However, once this higher concern is reached, all are better off; the customer, the worker, the manager, and the owner.[4]

In the events given in the preceding paragraphs, the manager was confronted with the areas of concern listed previously. Take another example; a maintenance supervisor of an FBO relates his experience:

GOVERNMENT REGULATIONS

Controlling costs in aviation maintenance is one of the most serious challenges facing the industry today. I have interviewed many experienced maintenance men and have arrived at three conclusions: 1) Government regulations, particularly the FAA, have pervaded the industry down to the point where it is nearly impossible to cut corners to save a dollar. For example, regulations permit a $5.00 an hour A&P to perform engine maintenance such as removing a cylinder head at will, but not allowing a half million dollar service and repair facility to perform the act unless they outline the operation as part of their FAA operating certificate, provide for the inspection of etc., etc., etc. The paperwork costs more than the job! Every task is spelled out thus and so; 2) The constant changing of equipment has eliminated the opportunity to establish any standard data. A brake repair could take as little as one hour up to ten hours on the same type of aircraft. During one interview, two experienced men brought in a piece of material that had required the combined experience of three men to figure out how to fix. Most maintenance men refuse to even attempt an estimate of repairs because of the variables; 3) I was appalled at the trend within most service departments to pass on increased costs to the customer. The automotive industry has established a flat rate for most repairs but the average aviation facility seems to avoid this management tool. The answer most received from the question about improving the profit picture was, "Raise our Prices!" Not one man even considered ways to improve the efficiency of his shop or effectively cut costs. Labor performance reports were unheard of. About the only target for maintenance control was that 95% of the man-hours expended in maintenance should appear on customer work orders.

For the small aviation firm, the manager cannot possibly undertake the rigid studies used as management tools by larger and more sophisticated aviation companies as discussed subsequently.

Cost appraisal methods, essential analysis methods, equipment cost targets, rime indexes, indicators, etc., all appear to be management tools of complex and long established maintenance departments. In aviation, only the airlines, military, and large fleets of corporate aircraft have the capacity to establish such programs. An individual aircraft owner can, however, establish a maintenance budget based on the hours of use on his or her machine. Corporate flight managers and professional pilots have work sheets to assist them in the analysis of maintenance and total operating expenses. These manufacturer-provided items can serve to express targets and if current expenses are properly maintained, they will ultimately reflect the use or abuse by the respective operator.

For example, pilots continually operating out of short airstrips can expect rapid brake wear. The target cost, the budget or "Dream Sheet," has proven to be a great value to a profit-oriented company. If a company is interested in reducing costs, it must initiate a maintenance cost reduction program commensurate with the complexity and cost of its equipment. If it continues to ignore maintenance costs and continues to add dollars to the customers' bills, it will soon be out of business. Unfortunately, that is the undoing of many aviation FBOs.

These examples point out a few of the real-world problems the manager deals with in his or her everyday job; but it is interesting to note that these types of illustrations are never found in the typical text on business management.

A number of managers get concerned about inventory control. Usually, there are three well-used methods that will suffice in getting accurate inventory data: 1) review by observation, 2) actual count or physical check, and 3) perpetual or "running" inventory. If the company wants to get more control and sophistication in its inventory system, then it will be concerned about "economic order quantity" (EOQ). Briefly, the EOQ is the best purchase price for a given lot at a specified time. Following EOQ purchases as much as realistically possible will gain a lower purchase price at a continuous rate and will usually allow for more stock room space.

Quality assurance is certainly one of management's biggest headaches. Poor product quality or workmanship quality results in the end in an unhappy customer or perhaps a dead one, since aviation flight accidents have caused fatalities that were the direct result of poor maintenance practices. Unfortunately for the company and the manager, there is almost always a legal suit that follows this type of accident. Fortunately, on the other hand, good quality assurance practices can be instilled in every worker by providing incentives, introducing proper training, adhering to manufacturer, FAA, and company guidelines and checklists, and by assuring good, rigid inspection procedures.

Maintenance Management Problem Areas

To establish and enforce standards of continued airworthiness, the FAA has maintenance inspectors[5] checking on the maintenance performed by aviation maintenance facilities to ensure these organizations comply with established standards. To understand the inspection procedure is to know what must be inspected, when it is inspected, and what the inspector is looking for.

There are, normally, two types of maintenance shops. The first and most common is the noncertificated shop. The other is the certified repair station (FAR Part 145). A noncertificated shop may be a one-person operation or may employ several certified A&P mechanics and noncertified helpers. If it is a one-person shop, that individual must hold the proper FAA ratings. Certificated repair stations come in a variety of classifications (see Chapter 6 for the different classifications). The certificated repair station must also meet certain requirements for housing, facilities, and personnel, with which the noncertified shop does not have to comply. A certified shop must maintain a complete records history of supervisory and inspection personnel, including the people responsible for management and those responsible for technical supervision. A certificated repair station may employ a combination of certified mechanics and repairmen, and noncertified helpers.

When an FAA inspector visits a certificated repair station, he or she will be checking to see the personnel working on aircraft understand their respective jobs, and that the appropriate technical specifications, airworthiness directives, service bulletins, and other mandatory documents are being used. The shop must also have the equipment recommended by the manufacturer (or its equivalent) for work on those aircraft.

A maintenance manager may not know when an inspector will stop by to check on his or her operation (however, many inspections are scheduled in advance). Whether the inspector announces his or her visit in advance or makes a no-notice inspection visit depends on the complaint record of the station. A public complaint may also prompt an inspection. If the FAA receives a formal complaint against a repair facility, an inspector must look into it. Maintenance inspectors will also come in for a spot check inspection. In a spot check, the entire operation may be checked, or just one or two areas that may have been lacking in a previous inspection. These inspections are usually unannounced.

One thing that will help a repair station look good on inspections is to know what the inspector is looking for. In an inspector's own words: "I will often look for an airplane that has just come out of an annual inspection to see how well the inspection was conducted. If I find an inspection plate where I can tell from a fastener or dirt around it that it hasn't been removed, I'll know it wasn't inspected properly. You can't inspect it if you can't see it."

If an inspector is performing an overall inspection, he or she will usually want to see the certificate of each employee who holds one. If the inspector is observing a mechanic at work, he or she will review the job being worked on to determine if the mechanic knows what he or she is doing. If the job requires FAA or manufacturer approved technical data, the inspector will check to see if the data is readily available to the person performing the maintenance. If the inspection is being performed as a result of an aircraft accident, the inspector will go to the repair station that performed the maintenance on that aircraft and check all records pertaining to work on that aircraft and will go over them very carefully. If a manager in charge of maintenance has competent people working for him or her who follow the proper technical data and airworthiness directives, he or she has no reason to fear a visit from an FAA inspector. Thus, the best defense against an FAA inspector is a strong in-house quality control and inspection program. In this way the repair station can find its own weak spots and correct them before the inspector discovers them. If a repair station fails to maintain the prescribed standards an inspection can have it rendered inoperational for up to one year.

Administrative concerns is an area normally dreaded by maintenance managers because they know that a fair portion of each work day must be put to the task of administrative paper work. Perhaps the most important task is to assure that all records are complete regarding FAA requirements in the accomplishments of all aircraft maintenance. Managers must also concern themselves with personnel records and filing pertaining to such items as time cards, job order requirements, parts stocking and control, and salary records. Government regulations, advisory circulars, FAR Part changes, and many other guidance and regulatory documents must be reviewed, brought to the attention of the workers, and filed in appropriate binders in the company library. Managers are totally responsible for their own maintenance organizations, assuring efficient systems through the development and administration of written policies, procedures, and requirements.

As a company grows, so does the maintenance department. A maintenance manager will start looking for assistance due to the added administrative burdens and will, no doubt, be making more use of the company's computer. The key managerial concerns are cost control and profitability. Good managers can reduce their concerns by increasing their knowledge through educational courses in administration. In today's environment, everyone has ample opportunity to take almost any type of course desired in nearly any geographical area. Business schools and universities with satellite campuses are widely scattered, so there is no excuse for not attending.

Although some managers seem to have no problem keeping a business thriving, they are the exceptions. To qualify as one of these exceptions a manager needs to be imaginative, forward thinking, original, flexible, quick

Maintenance Management Problem Areas

thinking, and just a little lucky. Even if a manager is not the things listed above, there are a few things he or she *should not do* to help ensure his or her business survives the "real world" economic cycles.

In the 1970s, even though the economy was in better financial shape than it is now, the University of Pittsburgh's Bureau of Business Research conducted a detailed study to try to pinpoint why businesses fail. According to the authors of the report, red ink is an indication, not a cause, for a breakdown in a company's health.[6] Today, when not only small FBOs but a few commuters and major airlines are on the brink of bankruptcy, the results of that study are more important than ever. Following are the 10 major failings of management (as outlined in the study) that were contributing factors, if not the primary factors, that caused a profitable business to founder: (1) keeping inadequate records, (2) ignoring new developments in one's field, (3) incurring cumulative losses, (4) relying on one account or customer, (5) being one's own expert, (6) building a family empire, (7) forgetting about cost analysis, (8) ignoring a competitor's mistakes, (9) expanding beyond one's resources, and (10) management's going off in different directions.

Maintenance management concerns are severe and for many problems there are no quick fixes. The manager's job is demanding and complex, and the tasks are not easy. However, if he or she applies himself or herself diligently, establishes good human relations with personnel, develops an effective organizational concept and maintenance procedures, he or she will undoubtedly find his or her concerns minimized. It is a basic assumption that the manager knows the maintenance field; so, therefore, he or she must learn to bridge the gap in those areas where he or she has limited knowledge or background.

13

Forecast and Summary

THE PREVIOUS CHAPTERS HAVE DWELT on the role of maintenance and its managers in the aviation field. The aviation maintenance employee is a highly skilled worker whose contribution to the industry cannot be overestimated. As equipment and aircraft become more complex and the industry continues to grow, the maintenance function will become more important and the position requirements more specialized.[1] Aviation maintenance mechanics, already well trained, will be required to be even better trained to work in more complex specialty areas.

To give the reader an indication of the growth of aviation in the United States, consider the data in Table 9. These figures point to a growing need for more skilled personnel in the aviation field in the years to come. The general aviation fleet alone has almost doubled in a 10-year period (1971–81).

From the passing curiosity it encountered 79 years ago, aviation today is viewed increasingly as a necessary means in the nation's travel requirements. General aviation is also becoming increasingly sophisticated with more than half of all operations having a business orientation. No more considered a luxury, aviation is now subject to the same pressures as other major industries. Diverse policies and environmental and economic conditions increasingly affect action taken by the aviation activity. Many factors exist which make the future of aviation uncertain. Not the least of these is the

Forecast and Summary

TABLE 9
GROWTH OF UNITED STATES AVIATION

	Passengers Carried	Aircraft Miles Flown (000)	General Aviation Fleet	Air Carrier Fleet	Fatal Accidents/ 100,000 Departures/ Air Carriers
1938	1,475,118	76,136			.092
1971	173,664,737	2,343,578	133,537	2,389	.121
1980	296,903,000	2,763,817	213,012	2,712	.000
1981	285,720,000		217,108	2,808	.076

Source: National Transportation Safety Board (NTSB), News Digest (1983).

impact of the Arab world on the cost and availability of fuel. Added to the fuel problem is the economic concern that many nations throughout the world have today. Aviation has a remarkable record for meeting challenges; its growth to date is an example. So the question is really not whether the aviation industry will grow but what will be its annual growth rate?

One of the prime concerns confronting the airlines today is the residual effects created by the Airline Deregulation Act of 1978. The act has presented major problems to many airlines, and they have attempted to meet the challenge through eliminating many unprofitable short-haul routes, freezing purchases of new aircraft, all but abandoning the requirement for new wide-bodies, contracting with commuters for a propping up of their terminal connections, and meeting the competition with many variations of ticket price reductions. The effect of deregulation on smaller communities can already be seen. Major carriers are suspending operations into a number of small municipalities, but this is proving a boon for the commuter airlines. These airlines are finding significant opportunities, and their growth rates are impressive. The commuters are also benefiting by the support of Congress in extending the equipment loan guarantee to them. Furthermore, the subsidy program for smaller communities now has become available to these airlines. These congressional actions along with the commuter trend of picking up existing markets from the major carriers and the growing demand for scheduled short-haul service should foster growth and stability among the commuter airlines over the long term.

The Airline Deregulation Act in 1978 was and still is an important event that will affect the development of aviation in the United States. For one thing, it diminishes the federal role in the commercial aspects of aviation. For another, it will mean that factors such as increased competition and higher costs will require more effective use of equipment, personnel, route planning, and greater cooperation, certainly between the majors and the commuters.

The largest sector of aviation in terms of flight operations is general aviation. With the exception of the commuters, deregulation has no direct effect on these operations. However, due to the development of improved and new reliever airports, both the business and personal flight operations will continue to grow. Federal concern for general aviation and the other sectors of aviation is shown in the efforts to improve safety throughout the national airspace system.

The recent surge in sales over the past several years of turboprop aircraft, primarily to corporate and commuter airline customers, is due to the greater fuel efficiency of these aircraft in comparison to pure jets. Given relatively short runway lengths and distance between takeoffs and landings of corporate and commuter flights, turboprop aircraft appear to be an excellent compromise between speed and fuel costs. With the many uncertainties in the aviation industry today, it is not likely that the application of new technology in aviation will forge ahead as it has in the past. Cost, fuel efficiency, and environmental concerns will be responsible for whatever new technology is developed.

The aviation industry, while continuing to grow, is being subjected to more diversified policy-making forces. The federal government's role is becoming more limited, and environmental concern along with economic uncertainty is putting new pressure on the aviation network. Aviation managers must become more sensitive to economic and world events and use this caution in their day-by-day planning and forecasting.

One thing is certain: the role of the skilled maintenance mechanic will be in more demand than ever, if not in the major carriers' maintenance shops then certainly with the commuters or corporate personnel. Growth in employment among the skilled trades has exceeded historical trends over the past few years with women coming into the aviation labor force in significant numbers. Although the growth rate in employment hovered at 3.6 to 3.8 percent into the late 1970s, in future years it is expected to decline to 1.1 percent by 1991.[2]

A new concept in aviation safety regulations entitled Regulation by Objective (RBO)[3] is being considered by the FAA for future guidance in its policy-making edicts. Because of the ever-changing operating environment, the FAA is considering replacing traditional "how to" regulations with the safety objectives they are intended to achieve. This will allow the affected operators to assess their operations and seek more effective and efficient methods of complying with safety objectives while maintaining the highest level of safety. The action is consistent with Executive Order 12291 and the Regulatory Flexibility Act of 1980.

Forecast and Summary 163

Current regulations have been developed over the last 50 years in response to growth and change within the aviation industry. The earliest Federal Aviation Regulations, developed in the 1930s, covered the domestic operations of scheduled operators. In 1945, regulations to cover overseas operations ("flag" air carriers) and nonscheduled operations ("supplemental" and "nonscheduled" air carriers) were added. Over the next 35 years, detailed and complex regulations were necessitated by factors such as economic regulation by the Civil Aeronautics Board (CAB) and changes in the aircraft available for passenger-carrying operations. At various times during this period, federal regulations made distinctions based not only on the domestic, flag, and supplemental categories, but also on the takeoff weight of the airplane ("large" versus "small" airplanes), type of engine (reciprocating versus turbine), and number of passengers that could be carried.

Further distinctions, based on the size and economics of operations, led to additional categories of operators such as commercial operators, air taxi operators, commuter operators, and others. As the variety of aircraft and variety of operators grew, so did the complexity of the regulations. Much of this complexity resulted because airlines were treated differently depending on whether they required economic authority from the CAB and, if so, depending on the type of that authority. Some of this historical development was related to the fact that the CAB, for many years, had both economic and safety regulatory authority. Many of the distinctions that led to the current set of regulations are no longer valid.

For many years there were few major complaints from either the aviation community or the affected public about federal aviation safety regulations. One indication of this fact is that aviation safety regulations have seldom been the subject of litigation. Before 1970, federal regulations in general were rarely challenged successfully. However, throughout the 1970s regulatory litigation flourished and agencies such as the Environmental Protection Agency and the Occupational Safety and Health Administration could expect to be challenged, often successfully, by the regulated industry, a public interest group, or both, on virtually every significant regulatory document issued.

Aviation safety regulations have not been similarly challenged perhaps because the complex "how to" regulations that grew over the last 40 years are serving the best interests of the regulated industry and the public, or perhaps because the aviation industry became accustomed to economic regulations. Whatever the reasons, it is now clear that change in the economic regulatory climate has affected the attitude of regulated aircraft operators. Some of the newer, expanding operators are prone to question the appropriateness of "how to" regulations and are inclined to seek better, more efficient ways of achieving safety objectives. The innovative approaches to solving long-time safety objectives have been initiated by the newer and smaller operators should not be surprising. It has been estimated that nearly half of the technological innovations introduced in the United States between 1953 and 1973 were developed by businesses with fewer than 1,000 employees.

In any case, a regulatory system that sets forth complex and specific methods intended for general applicability throughout an industry tends to inhibit innovation within that industry. A number of FAA regulations are already written, at least partially, in objective terms that allow for individual approaches subject to FAA approval (the continuous airworthiness maintenance program is one example).

What the future holds in this regulatory movement is a moot question. Whatever the case, it is expected that there will be changes, some very extensive in nature. This is what Paul R. Ignatius, President and Chief Executive Officer, Air Transport Association, has to say about the future of air transportation.

> Expanded capacity and technological gains are needed to meet airline passenger cargo growth in the next 10 years. This will be achieved essentially with the same-sized fleet of about 3,000 airline aircraft. General aviation is expected to expand from 200,000 to more than 300,000 aircraft.
>
> Growth and reliability have been the hallmark of U.S. air transportation since its start in 1926. By 1991, according to the FAA, the airlines will be carrying more than 450 million passengers a year and transporting 12 billion ton miles of cargo. The nation in the years ahead will continue to rely heavily on the national air transportation system and its people.[4]

Appendixes
Notes
Index

Appendix A
FAA Advisory Circulars—Maintenance (Abbreviated Listing)

AC Number	Subject
00-2VV	Advisory Circular Checklist (4-15-82). Transmits the revised checklist of current FAA advisory circulars (ACs).
00-44R	Status of Federal Aviation Regulations (2-1-82). Sets forth the current publication status of the Federal Aviation Regulations, any changes issued to date, and provides a price list and ordering instructions.
20-77	Use of Manufacturers' Maintenance Manuals (3-22-72). Informs owners and operators about the usefulness of manufacturers' maintenance manuals for servicing, repairing, and maintaining aircraft, engines, and propellers.
20-97	High-Speed Tire Maintenance and Operational Practices (1-38-77).

AC Number	Subject
20-109	Service Difficulty Program (General Aviation) (1-8-79).
	Describes the Service Difficulty Program as it applies to general aviation activities and provides instructions for completion of the Malfunction or Defect Report (M or D), FAA Form 8010-4.
21-9A	Manufacturers Reporting Failures, Malfunctions, or Defects (5-26-82).
	Provides information to assist manufacturers of aeronautical products (aircraft, aircraft engines, propellers, appliances, and parts) in notifying the Federal Aviation Administration of certain failures, malfunctions, or defects resulting from design or quality control problems in the products which they manufacture.
21-19	Installation of Used Engines in New Production Aircraft (4-26-82).
	Advises that under certain specified criteria, used engines may be used in new production aircraft.
39-6H	Summary of Airworthiness Directives (4-12-82).
	Announces the availability of a new summary of Airworthiness Directives (AD), dated January 1, 1982, and provides information for ordering these publications.
43-3	Non-destructive Testing in Aircraft (5-11-74).
	Reviews the basic principles underlying nondestructive testing.
43-4	Corrosion Control for Aircraft (5-11-73).
	Summarizes current available data regarding identification and treatment of corrosive attack on aircraft structure and engine materials.
43-4 (chg. 1) (3-1-74)	Provides additional information on identification and treatment of corrosion attack on aircraft structures. Adds a new Chapter 14—Corrosion Control of Aircraft Used in Agricultural Cropdusting Operations.

AC Number	Subject
	Clarifies the discussion on the removal of corrosion and treatment of corroded areas.
43-5	Airworthiness Directives for General Aviaiton Aircraft (8-13-74).
	Points to areas of misunderstanding regarding: aircraft owners' and operators' responsibility for complying with regard to performance of ADs, and maintenance records entries for ADs required by FAR 91.173(a)(2)(v) and FAR 43.9.
43-6A	Automatic Pressure Altitude Encoding Systems and Transponders Maintenance and Inspection Practices (11-11-77).
	Provides information on the installation of encoding altimeters based upon recently acquired operating experience and on the maintenance of ATC transponders.
43-7	Ultrasonic Testing for Aircraft (9-24-74).
	Describes methods used in ultrasonic nondestructive testing, discusses the many advantages, and points out the simplicity of the tests. Contains many illustrations.
43-9A	Maintenance Records: General Aviation Aircraft (9-9-77).
	Provides information to assist maintenance personnel in fulfilling their responsibility under FAR Section 43.9.
43-10	Mechanical Work Performed on U.S. and Canadian Registered Aircraft (1-26-76).
	Provides information and guidance to aircraft owners/operators and maintenance personnel concerning mechanical work performed on U.S. registered aircraft by Canadian maintenance personnel and on Canadian registered aircraft by U.S. maintenance personnel.

AC Number	Subject
43-11	Reciprocating Engine Overhaul Terminology and Standards (4-7-76).
	Discusses engine overhaul terminology and standards that are used by the aviation industry.
43-12	Preventive Maintenance (7-16-76).
	Provides information concerning preventive maintenance and who may perform it.
43-15	Recommended Guidelines for Instrument Shops (8-15-77).
	Provides guidelines concerning environmental conditions for instrument repair and overhaul shops and information on calibration of test equipment.
43-9-1C	Instructions for Completion of FAA Form 337 (12-20-73).
	Provides instructions for completing revised FAA Form 337, major Repair and Alteration (Airframes, Powerplant, Propeller, or Appliance).
43.13-1A	Acceptable Methods, Techniques and Practices—Aircraft Inspection and Repair (4-17-72)
	Contains methods, techniques, and practices acceptable to the Administrator for inspection and repair to civil aircraft. Published in 1973.
43.13-1A (chg. 1) (5-12-75)	Transmits new and revised material for basic advisory circular.
43.13-1A (chg. 2) (12-22-76)	Transmits revised material concerning aircraft instrument adjustment.
43-13-2A	Acceptable Methods, Techniques, and Practices—Aircraft Alterations (6-9-77).
	Contains methods, techniques, and practices acceptable to the Administrator for use in altering civil aircraft.

FAA Advisory Circulars—Maintenance

AC Number	Subject
43-203A	Altimeter and Static System Tests and Inspections (6-6-67).
	Specifies acceptable methods for testing altimeters and static system. Also provides general information on test equipment used and precautions to be taken.
65-2D	Airframe and Powerplant Mechanics Certification Guide (1-30-76).
	Provides information to prospective airframe and powerplant mechanics and other persons interested in FAA certification of aviation mechanics.
65-9A	Airframe and Powerplant Mechanics—General Handbook (4-12-76).
	Designed as a study manual for persons preparing for a mechanic certificate with airframe or powerplant ratings. Emphasis in this volume is on theory and methods of application, and is intended to provide basic information on principles, fundamentals, and airframe and powerplant ratings.
65-11A	Airframe and Powerplant Mechanics Certification Information (4-21-71).
	Provides answers to questions most frequently asked about Federal Aviation Administration certification of aviation mechanics.
65-12A	Airframe and Powerplant Mechanics Powerplant Handbook (4-12-76).
	Designed to familiarize student mechanics with the construction, theory of operation, and maintenance of aircraft powerplants.
65-13C	FAA Inspection Authorization Directory (10-19-77).
	Provides a new directory of all FAA certificated mechanics who hold an inspection authorization as of August 31, 1977.
65-15A	Airframe and Powerplant Mechanics Airframe Handbook (4-12-76).

AC Number	Subject
	Designed to familiarize student mechanics with airframe construction, repair, and the operating theory of the airframe system.
65-18	Report Availability of a Survey of the Aviation Mechanics Occupation (9-4-74).
	Announces the public availability of the 1974 report on a Survey of the Aviation Mechanics Occupation.
65-19A	Inspection Authorization Study Guide (11-17-76).
	Provides guidance for persons who conduct annual and progressive inspections and approve major repairs and/or alterations of aircraft. It stresses the importance that certificated mechanics, holding IAs, have in air safety. Primarily intended for mechanics who hold or are preparing to take the test for an inspection authorization.
91-26	Maintenance and Handling of Air-Driven Gyroscopic Instruments (10-29-69).
	Advises operators of general aviation aircraft of the need for proper maintenance of air-driven gyroscopic instruments and associated air filters.
91-59	Inspection and Care of General Aviation Aircraft Exhaust Systems (8-20-82).
	Emphasizes the safety hazards of poorly maintained single-engine aircraft exhaust systems (reciprocating powerplants) and highlights points at which exhaust system failures occur. Further, it provides information on the kind of problems to be expected and recommends the performance of ongoing preventive-maintenance and maintenance by pilots and mechanics respectively.
120-16B	Continuous Airworthiness Maintenance Programs (9-14-78).
	Provides air carriers and commercial operators with guidance and information pertinent to certain provisions of Federal Aviation Regulations Parts 121 and 127 regarding continuous airworthiness maintenance programs.

FAA Advisory Circulars—Maintenance

AC Number	Subject
120-17A	Maintenance Control by Reliability Methods (3-27-78). Provides information and guidance materials which may be used to design or develop maintenance reliability programs utilizing reliability control methods.
121-1A	Standard Operations Specifications—Aircraft Maintenance Handbook (6-26-73). Provides procedures acceptable to the Federal Aviation Administration which may be used by operators when establishing inspection intervals and overhaul times.
121-1A (chg. 1) (1-23-75)	Updates the overhaul and inspection/check period of selected airframes, powerplants, propellers, and appliances in relation to current industry standards.
121-22	Maintenance Review Board (MRB) (1-12-77). Provides guidelines for establishing and conducting an MRB on newly manufactured aircraft, powerplant, or appliance to be used in air carrier service.
129-3	Foreign Air Carrier Security (6-10-82). Provides information and guidance for the implementation of Change 7 to Federal Aviation Regulations (FAR) Part 129 (Sections 129.25, 129.26, and 129.27), which became effective September 11, 1981. The sensitive nature of many security procedures precludes full discussion in this circular. A list of Civil Aviation Security Offices and Principal Security Inspectors for foreign air carriers is included as Appendix 1.
135-5A	Maintenance Program Approval for Carry-on Oxygen Equipment for Medical Purposes (11-12-76). Provides a means whereby air taxi operators may submit a maintenance program to comply with FAR Part 135, Section 135.114.
137-7	FAR Part 135: Additional Maintenance Requirements for Aircraft Type Certificated for Nine or Less Passenger Seats (10-24-78).

AC Number	Subject
	Provides information on establishing methods for compliance with additional maintenance requirements of FAR Part 135, revised December 1, 1978, for certain air taxi operators and commercial operators (ATCO).
140-1J	Consolidated Listing of FAA Certificated Repair Stations (7-27-77).
	Provides a revised directory of all FAA certificated repair stations as of May 31, 1977.
140-2L	List of Certificated Pilot Schools (5-11-78).
	Provides a list of FAA certificated pilot flight and ground schools as of March 1978.
140-6	The Development and Use of Major Repair Data Under Provisions of Special Federal Aviation Regulation (SFAR) 36 (3-20-78).
	Provides information related to the issuance of an authorization to allow repair stations, air carriers, and air taxi/commercial operators of large aircraft to develop and use major repair data not approved by the FAA Administrator in accordance with the requirements of SFAR 36.
145.101-1A	Application for Air Agency Certificate Manufacturer's Maintenance Facility (3-12-77).
	Explains how to obtain a repair station certificate.
147-2S	Directory of FAA Certificated Aviation Maintenance Technician Schools (5-5-78).
	Provides a revised directory of all FAA certificated aviation maintenance technician schools as of March 1978.
147-3	Phase III, A National Study of the Aviation Mechanics Occupation (3-22-71).
	Announces the availability for purchase by the public of a report of Phase III, A National Study of the Aviation Mechanics Occupation.
147-4	Reports availability of a Study of Test Materials Used in Aviation Maintenance Technician Schools (9-3-74).

AC Number	Subject
	Announces the public availability of the 1974 report Study of Test Materials Used in Aviation Maintenance Technician Schools.
183-32C	Federal Aviation Administration Designated Maintenance Technician Examiners Directory (6-24-82). Transmits a consolidated directory of DMEs (Designated Mechanic Examiners) (Appendix 1) and DPREs (Designated Parachute Rigger Examiners) (Appendix 2), as of June 8, 1982, under the authority of FAR Part 183.

Appendix B
Reference to FAA Reporting Requirements and Forms

A. This section provides an index and quick reference to the various forms and reporting requirements contained in Order 8600.1, "General Aviation Airworthiness Inspector's Handbook." It is intended to provide guidance to the proper form and/or report when performing general aviation airworthiness functions.

TITLE	MEDIUM	REFERENCE
1. Aircraft/Part Identification and Release Tag	FAA Form 8020-2	Chapter 2, Section 6
2. Report of Maintenance/Avionics Seminars and Clinics	Narrative	Chapter 2, Section 7
3. Industry Compliance With FAR Effective Dates	Narrative	Chapter 2, Section 11
4. Certificate, Authorization/Designation Action Request	AC Form 8300-10 (Figure 2-3)	Chapter 2, Sections 12, 13

Title	Medium	Reference
5. Inspection and Surveillance Record (Fueling Facilities Surveillance)	FAA Form 3112	Chapter 2, Section 16
6. Maintenance Type Certification Activity Summary	Narrative	Chapter 3, Section 1
7. Application For Type Certificate, Production Certificate or Supplemental Type Certificate	FAA Form 8110-12	Chapter 3, Section 2
8. Major repair and Alteration (Airframe, Powerplant, Propeller or Appliance)	FAA Form 337	Chapter 3, Sections 3, 4; Chapter 4, Section 8; Chapter 6, Sections 1, 9
9. Standard Airworthiness Certificate	FAA Form 8100-2	Chapter 3, Section 6
10. Service Difficulty Report	FAA Form 8070-1	Chapter 3, Section 11; Chapter 6, Section 1
11. Malfunction or Defect Report	FAA Form 8010-4	Chapter 3, Sections 11, 18; Chapter 6, Section 1
12. Aircraft Condition Notice	FAA Form 8620-1 (Figure 3-1)	Chapter 3, Section 11; Chapter 4, Sections 12, 13; Chapter 6, Section 1
13. Report of Aircraft Surveillance	FAA Form 3112	Chapter 3, Section 11

FAA Reporting Requirements and Forms

Title	Medium	Reference
14. Follow-up of Unairworthy Aircraft	Narrative	Chapter 3, Section 15
15. FAA Inspection Reminder	FAA Form 8600.1	Chapter 3, Section 15
16. Operations Specifications	FAA Form 1014	Chapter 3, Section 20; Chapter 6, Section 11
17. Application For Airworthiness Certificate	FAA Form 8130-6	Chapter 3, Section 23
18. Application For Repair Station Certificate and/or Rating	FAA Form 8310-3 (Figure 4-1)	Chapter 4, Sections 1, 3, 5
19. Air Agency Certificate (Repair Station)	FAA Form 8000-4 (Figures 4-2, 4-2A, 4-2B)	Chapter 4, Sections 1, 3
20. Repair Station Operations Specifications	FAA Form 8000-4-1 (Figures 4-3, 4-4)	Chapter 4, Sections 2, 3
21. Certification, Authorization/Designation Action Request (Repair Station Certificate)	AC Form 8300-10 (Figure 2-3)	Chapter 4, Section 3
22. Application For Repair Station Certificate and/or Rating (Record of Action)	FAA Form 8310-3	Chapter 4, Section 5
23. Letter of Application (Manufacturer's Maintenance Facility)	Narrative	Chapter 4, Section 9
24. Air Agency Certificate (Manufacturer's Maintenance Facility)	FAA Form 8000-4 (Figure 4-5)	Chapter 4, Section 9

Title	Medium	Reference
25. Certification, Authorization/Designation Action Request (Manufacturer's Maintenance Facility Certification)	AC Form 8300-10 (Figure 2-3)	Chapter 4, Section 9
26. Inspection and Surveillance Record (Repair Station Inspection)	FAA Form 3112	Chapter 4, Section 11
27. Malfunction or Defect Report (Repair Station Reports)	FAA Form 8010-4	Chapter 4, Sections 11, 12
28. Inspection and Surveillance Record (FAA Maintenance Facility)	FAA Form 3112	Chapter 4, Section 11
29. Application For Pilot School Certificate	FAA Form 8420-8	Chapter 4, Section 13
30. Report of Aircraft Surveillance (Pilot Schools)	FAA Form 3112	Chapter 4, Section 13
31. Malfunction and Defect Report (Pilot School Aircraft)	FAA Form 8010-4	Chapter 4, Section 13
32. Application for Repair Station Certificate and/or Parachute Loft	FAA Form 8310-3	Chapter 4, Section 14
33. Air Agency Certificate (Parachute Loft)	FAA Form 8000-4 (Figure 4-6)	Chapter 4, Section 14
34. Certification, Authorization/Designation Action Request (Parachute Loft Certification)	AC Form 8300-10 (Figure 2-3)	Chapter 4, Section 14

FAA Reporting Requirements and Forms

Title	Medium	Reference
35. Major Repair and Alteration (Airframe, Powerplant, Propeller or Appliance). Parachute Alteration Date	FAA Form 337	Chapter 4, Section 15
36. Application for Air Taxi Commercial Operator Certificate (Inspection Results)	FAA Form 8000-6 (Figures 4-7, 4-8)	Chapter 4, Section 17; Chapter 6, Section 1
37. Operations Specifications (Aircraft Maintenance)	FAA Form 1014	Chapter 4, Section 17
38. Aviation Safety Inspector's Credential	FAA Form 110A (Figures 4-9, 4-10)	Chapter 4, Section 19
39. Application for Identification or Credential Card	FAA Form 1600-11 (Figures 4-11, 4-12)	Chapter 4, Section 19
40. Request for Access to Aircraft	FAA Form 8430-13 (Figure 4-13)	Chapter 4, Section 19
41. Inspection and Surveillance Record (Enroute Inspection)	FAA Form 3112 (Figure 4-14)	Chapter 4, Section 19
42. Operations Specifications (Weight and Balance Control)	FAA Form 1014 (Figure 4-15)	Chapter 4, Section 20
43. Aviation Maintenance Technician School Certification and Ratings Application	FAA Form 8310-6 (Front) (Figure 4-18)	Chapter 4, Section 21

Title	Medium	Reference
44. Certification, Authorization/Designation Action Request (Aviation Maintenance Technician School Certification)	AC Form 8300-10 (Figure 2-3)	Chapter 4, Section 21
45. Aviation Maintenance Technician School Inspection Report	FAA Form 8310-6 (Reverse) (Figure 4-19)	Chapter 4, Sections 22, 25
46. Air Agency Certificate (Aviation Maintenance Technician School Certification)	FAA Form 8000-4 (Figure 4-20)	Chapter 4, Section 22
47. Aviation Maintenance Technician School Certification and Ratings Application (Change of Location)	FAA Form 8310-6	Chapter 4, Sections 24, 25
48. Aviation Maintenance Technician School Norms Vs. National Passing Norms	Computer Run (Figures 4-21, 4-22, 4-23)	Chapter 4, Section 25
49. Airman Written Test Application	AC Form 8080-3	Chapter 4, Section 25
50. GADO/FSDO Aviation Maintenance Technician School Norms Vs. National Passing Norms (Summary)	Computer Run (Figure 4-24)	Chapter 4, Section 25

FAA Reporting Requirements and Forms

Title	Medium	Reference
51. Aviation Maintenance Technician School Norms Vs. National Passing Norms (Summary)	Computer Run (Figure 4-25)	Chapter 4, Section 25
52. GADO/FSDO Aviation Maintenance Technician School Norms Vs. National Passing Norms (Summary)	Computer Run (Figure 4-26)	Chapter 4, Section 25
53. Aviation Mechanic Test Applicant Listing	Computer Run (Figure 4-27)	Chapter 4, Section 25
54. Aviation Mechanic General Test (Test Outline)	AC Form 8080-2-15	Chapter 4, Section 25
55. Aviation Mechanic Airframe Test (Test Outline)	AC Form 8080-2-16	Chapter 4, Section 25
56. Aviation Mechanic Powerplant Test (Test Outline)	AC Form 8080-2-17	Chapter 4, Section 25
57. Certification, Authorization/Designation Action Request (Change in School Status)	AC Form 8300-10 (Figure 2-3)	Chapter 4, Section 25
58. Airman Written Test Report	AC Form 8080-2	Chapter 5, Sections 1, 10, 11
59. Military Service Record	DD 214-205	Chapter 5, Section 1
60. Airman's Authorization For Written Test	FAA Form 8060-7	Chapter 5, Sections 1, 11
61. Airman Written Test Application	AC Form 8080-3	Chapter 5, Sections 1, 11

Title	Medium	Reference
62. Airman Certificate and/or Rating Application	FAA Form 8610-2 (Front) (Figures 5-1, 5-2)	Chapter 5, Sections 1, 8, 11
63. Airman Certificate and/or Rating Application (Privacy Act)	FAA Form 8610-2 (Supplemental) (Figure 5-5)	Chapter 5, Sections 1, 8
64. Airman Certificate and/or Rating Application (Results of Oral and Practical Tests)	FAA Form 8610-2 (Reverse) (Figures 5-3, 5-4)	Chapter 5, Section 1
65. Temporary Airman Certificate	FAA Form 8060-4 (Figure 5-6)	Chapter 5, Sections 1, 8, 10, 11
66. Airman Certificate	AC Form 8060-1	Chapter 5, Sections 1, 8, 11
67. Letter of Aeronautical Competency	Form Letter (Figure 5-6A)	Chapter 5, Section 1
68. Mechanic Application For Inspection Authorization	FAA Form 8610-1 (Figure 5-7)	Chapter 5, Sections 5, 6
69. Notice of Disapproval of Application	FAA Form 8060-5 (Figure 5-8)	Chapter 5, Sections 5, 11
70. Inspection Authorization	FAA Form 8310-5 (Figures 5-9, 5-9A)	Chapter 5, Section 5
71. Certification, Authorization/Destination Action Request (Issuance of IA)	AC Form 8300-10 (Figure 2-3)	Chapter 5, Section 5
72. Certification, Authorization/Designation Action Request (Failure to Renew IA)	AC Form 8300-10 (Figure 2-3)	Chapter 5, Section 6

FAA Reporting Requirements and Forms

Title	Medium	Reference
73. Statement of Qualifications (DMIR-DER-DPRE-DME)	FAA Form 8110-14 (Figures 5-12, 5-13)	Chapter 5, Section 10
74. Certificate of Authority	FAA Form 8430-9 (Figures 5-14, 5-15)	Chapter 5, Section 10
75. Certificate of Designation	FAA Form 8000-5 (Figure 5-16)	Chapter 5, Section 10
76. Certification, Authorization/Designation Action Request (Issuance of DME and DPRE)	AC Form 8300-10 (Figure 2-3)	Chapter 5, Section 10
77. Parachute Rigger Seal Symbol Assignment Card	FAA Form 3318 (Figure 5-19)	Chapter 5, Section 11
78. General Aviation Maintenance/Avionics Work Program and Activity Report	FAA Form 1380-16	Chapter 6, Section 1
79. Guide For Aircraft Maintainability Evaluation Summary	FAA Form 8320-15	Chapter 3, Section 1 and Appendix 4

B. This section provides an index and quick reference to the various reporting requirements and forms contained in Order 8320.12, "Air Carrier Airworthiness Inspector's Handbook."

Scheduled Reporting Requirements (2-16-78)

1. Air Carrier Aircraft/Engine Utilization Report (Monthly)	AC Form 8320-1	Chapter 3, Par. 760
2. List of Air Carrier Aircraft (Quarterly)	Narrative	Chapter 3, Par. 749

Appendix B

Title	Medium	Reference
Unscheduled Reporting Requirements (11-7-77)		
1. Air Carrier Maintenance Activities During Employee Strikes	Narrative	Chapter 3, Par. 798
2. Strike Surveillance	Dispatch Telephone	Chapter 3, Par. 798
3. Major Repair and Alterations	FAA Form 337	Chapter 6, Pars. 1757, 1764, 1790, 1830; Chapter 7, Par. 2345
4. Enroute Inspection	FAA Form 3112; SF-160; FAA Form 3430-13	Chapter 3, Pars. 696, 697
5. Ramp and Spot Checks	FAA Form 3112	Chapter 3, Par. 774
6. Repair Station Inspection	FAA Form 3112	Chapter 7, Par. 2391
7. Informal Surveillance Maintenance Activities	FAA Form 3112	Chapter 3, Par. 728
8. Proving Flight	FAA Form 3112	Chapter 3, Par. 458
9. Maintenance Facility Inspection	FAA Form 3112	Chapter 3, Par. 483
10. Surveillance of Maintenance Airman and Repairman	FAA Form 3112	Chapter 7, Par. 2740
11. Operations Specifications	FAA Form 1014	Chapter 3, Pars. 510, 518, 554
12. Air Travel Club Certification	Narrative	Chapter 4, Par. 1308

FAA Reporting Requirements and Forms

Title	Medium	Reference
13. Airman Certification	DD214 (Military); FAA Forms 8000-4, 8000-5, 8000-33, 8060-4, 8060-5, 8060-7; AC Forms 8080-2, 8080-3; FAA Forms 8110-14, 8310-1, 8610-2, 8310-5, 8430-9	Chapter 8, Pars. 2650, 2774, 2810, 2860, 2912, 2960
14. Temporary Grounding Air Carrier Aircraft	Narrative	Chapter 3, Par. 918
15. Special Report of Significant Failures, Malfunctions, and Defects on Wide-Bodied Jets	Narrative	Chapter 3, Par. 632
16. Maintenance Type Certification Activity Summary	Narrative	Chapter 2, Par. 100
17. Air Carrier Meeting Report/Attendees	Narrative	Chapter 3, Par. 668
18. Request for Mailing List Action	AC Form 8300-10	Chapter 2, Par. 252; Chapter 7, Par. 2444

Forms (11-7-77)

Title	Medium	Reference
1. Aviation Safety Inspector's Credential	FAA Form 110A	Chapter 3, Par. 692
2. Request for Access to Aircraft or Free Transportation	SF 160	Chapter 3, Par. 693
3. Armed Forces of the United States Report of Transfer or Discharge	DD214 (Military)	Chapter 8, Par. 2666

Title	Medium	Reference
4. Repair and Alteration	FAA Form 337	Chapter 6, Pars. 1757, 1764, 1790, 1830; Chapter 7, Par. 2345
5. Repair Station Operations Specifications	FAA Form 8000-4-1, Page 2	Chapter 7, Pars. 2226, 2369
6. Operations Specifications	FAA Form 1014	Chapter 3, Pars. 510, 515, 554, 622
7. Inspection and Surveillance Record	FAA Form 3112	Chapter 3, Pars. 458, 483, 696, 728, 774; Chapter 7, Par. 2391; Chapter 8, Par. 2740
8. Air Agency Certificate	FAA Form 8000-4	Chapter 7, Par. 2202, 2226
9. Certificate of Designation	FAA Form 8000-5	Chapter 8, Par. 2964
10. Airman Certification Privacy Act	FAA Form 8610-2	Chapter 8, Par. 2659
11. Airman Certificate	AC Form 8060-1	Chapter 8, Pars. 2666, 2777, 2778, 2818, 2920, 2922
12. Temporary Airman Certificate	FAA Form 8060-4	Chapter 8, Pars. 2662, 2665, 2666, 2667, 2778, 2917, 2918, 2922, 2965
13. Notice of Disapproval of Application	FAA Form 8060-5	Chapter 8, Pars. 2817, 2965

FAA Reporting Requirements and Forms   189

Title	Medium	Reference
14. Airman's Authorization for Written Test	FAA Form 8060-7	Chapter 8, Par. 2656
15. Airman Written Test Report	FAA Form 8080-2	Chapter 8, Pars. 2658, 2663, 2664, 2666, 2963
16. Airman Written Test Application	FAA Form 8080-3	Chapter 8, Par. 2656
17. Standard Airworthiness Certificate	FAA Form 8100-2	Chapter 6, Par. 1873
18. Statement of Qualifications	FAA Form 8110-14	Chapter 8, Pars. 2964, 2972
19. Special Airworthiness Certification	FAA Form 8130-7	Chapter 3, Par. 573
20. Certificate, Authorization or Designation Action Request	AC Form 8300-10	Chapter 7, Par. 2444
21. Mechanic's Application for Inspection Authorization	FAA Form 8610-1	Chapter 8, Pars. 2814, 2818, 2863
22. Airman Certificate and/or Rating Application	FAA Form *610-2	Chapter 8, Pars. 2659, 2660, 2662, 2663, 2664, 2666, 2667, 2777, 2778, 2916, 2922, 2965
23. Guide for Aircraft Maintainability Evaluation Summary	FAA Form 8320-15	Chapter 2, Par. 99(a), App. 4

Appendix C
Glossary

Air Cargo. Total volume of freight, mail, and express traffic transported by air. Statistics include the following: freight—commodities of all kinds (includes small package counter services); express—priority reserved freight and express services; and United States Mail—all classes of mail transported for the United States Postal Service.

Air Carrier. Any person or organization which undertakes, whether directly or indirectly, or by lease or any other arrangement, to hold itself out to the public to furnish common carriage transportation by aircraft.

Air Taxi. An air carrier certificated in accordance with FAR Part 135 and authorized to provide, on demand, public transportation of persons or property by aircraft.

Air Traffic Control. A service provided by the FAA to promote the safe, orderly, and expeditious flow of air traffic along the nation's airways and within all other controlled air space.

Air Transportation. The carriage of persons or property for compensation or hire, or the carriage of mail by aircraft in commerce.

Air Travel Club. An operator certificated in accordance with FAR Part 123 to engage in the carriage of members who are qualified for that carriage by payment of an assessment, dues, membership fees, or other similar remittance.

Alaska-Hawaii. Certificated route air carriers conducting scheduled operations within the states of Alaska or Hawaii.

All Cargo (418). All cargo (418) identifies an air carrier holding an "All Cargo Air Service Certificate," issued under Section 418 of the FAA Act, and certificated in accordance with FAR Part 121 to provide domestic air transportation of cargo, excluding service between any pair of points, both of which are in Alaska or Hawaii.

Aviation Gasoline. All piston-engine aircraft use a high-grade (high octane) gasoline as a fuel, in contrast to the jet fuel (generally kerosene) used in turbine-powered aircraft. This piston-engine fuel is called aviation gasoline—or AVGAS for short.

Basic Function. The one operation initially the basis for determining a deisgn.

Certificated Route Air Carrier. An air carrier certificated in accordance with FAR Parts 121, 127, or 135, which holds a certificate of public convenience and necessity issued by the Civil Aeronautics Board to conduct scheduled services over specified routes. These air carriers also provide nonscheduled or charter services as a secondary operation.

Charter Air Carrier. An air carrier holding a certificate of public convenience and necessity authorizing it to engage in charter air transportation.

Commerical Operators. This term encompasses a classification of operators who engage in contract operations involving the carriage by aircraft of persons or property for compensation or hire. Generally, the operation includes scheduled intrastate common carriage and/or private contract interstate services.

Commuter. An air carrier, certificated in accordance with FAR Part 135, authorized to provide air transportation of passengers or cargo pursuant to a published schedule of at least five round trips per week between two or more points, or to transport mail pursuant to a contract with the United States Postal Service.

Cost Reduction. Various actions conducted to reduce costs in all aspects of an organization or company. These actions are typically motivation programs, suggestion programs, time and motion studies, managerial decisions/edicts, PERT and other programming devices, management and employee improvement programs, etc.

Direct Maintenance (Expense). The cost of labor, materials, and outside ser-

Glossary

vices consumed directly in periodic maintenance operations and the maintenance, repair, or upkeep of airframes, aircraft engines, other flight equipment, and ground property and equipment.

Domestic Trunk Carrier. Certificated route air carriers operating primarily within and between the 50 states of the United States, over routes serving primarily the large communities.

"Eliminate Gold Plating." Same as Value Engineering. This phrase originated in the Office of the Secretary of Defense (OSD).

FBO. Fixed Base Operation or Fixed Base Operator; a sales and/or service facility located at an airport or the person who operates such a facility.

Federal Aviation Regulations (FARS). Mandatory requirements and standards issued by the FAA to govern all civil aviation activities.

Foreign-Flag Air Carrier. An air carrier other than a United States flag air carrier engaged in international air transportation.

Functional Analysis. The analysis of the functions performed by use of a value analysis study item. In general, this amounts to identifying all functions and determining the degree of relative importance of each function to the basic function, as well as the indicated costs needed to provide those functions.

Function. An operation (work) performed by a device or service.

General Aviation. All civil aircraft and aviation activity except that of the certified air carriers.

Helicopter. A type of aircraft that operates by means of revolving rotors or blades engine-driven about an approximately vertical axis. A helicopter does not have conventional fixed wings nor in any but some earlier models is it provided with a conventional propeller. Forward thrust and lift are furnished by the rotors. The powered rotor blades also enable the machine to hover and to land and take off vertically.

Helicopter Carriers. Domestic certificated route air carriers primarily employing helicopter aircraft in their operations.

Indirect Maintenance (Expense). Overhead or general expenses of activities involved in the repair and upkeep of property and equipment, including inspections of equipment in accordance with prescribed operational standards. Also included under this definition are expenses related to the administration of maintenance stocks and stores, the keeping of maintenance operations, records, and the scheduling, controlling, planning, and supervision of maintenance operations.

International/Territorial Carrier. A certificated route air carrier operating between the United States and its territories and possessions and foreign countries.

Jet Fuel. Practically all United States airlines use kerosene as fuel in their turbine-powered aircraft. This kerosene is commonly referred to as jet fuel, in contract to the aviation gasoline used in piston-engine aircraft.

Large Regional Airline. An airline that earns between $10 million and $75 million per year in revenues. Includes such airlines as PBA, Rio, Britt, and Mid Pacific. These airlines are mostly former commuter airlines or are new entrants.

Load Factor. The percentage of seating or freight capacity which is utilized.

Local Service Carriers. Domestic certificated route air carriers operating routes of lesser density between the smaller traffic centers and between those centers and principal centers, as opposed to domestic trunk carriers.

Maintenance. Includes inspection, repair, overhaul, preservation, and the replacement of parts, but not preventive maintenance.

Maintenance Instructions. Includes service bulletins, letters, or other publications concerning maintenance applicable to specified models and configurations (modification status or other groupings that influence maintenance needs).

Major Airline. An airline that earns $1.0 billion or more in annual revenues. Includes such airlines as United, Eastern, American, Delta, and Trans World. Mostly former trunk airlines are in this category although U.S. Air and Republic, which are former local service airlines, are also majors.

Medium Regional Airline. An airline that earns less than $10 million in annual revenues.

National Airline. An airline that earns between $75 million and $1.0 billion in annual revenues. This definition includes such airlines as Ozark, Pacific Southwest, Southwest, Hawaiian, and Alaska. Former local service and instrastate airlines make up this category of airlines, although a number of former supplemental airlines and new entrant airlines are also included.

New Entrant. An airline that has begun service since October 24, 1978, the effective date of the Airline Deregulation Act of 1978.

Nonscheduled Service. Revenue flights that are not operated in regularly scheduled service, such as charter flights.

Piston Aircraft. Aircraft fitted with the reciprocating engine in which pistons, moving back and forth, work upon a crankshaft or other device to create rotational movement of the propeller.

Preventive Maintenance. Simple, minor, preservation or replacement.

Regular Body Aircraft. A generic and commonly used term applied to jet aircraft, especially turbofans with a fuselage diameter of less than 200 inches and whose per engine thrust is less than 30,000 pounds (i.e., Boeing 707, 727; McDonnell Douglas DC-8, DC-9, etc.).

Scheduled Service. Transport service operated over an air carrier's certificated routes, based on published flight schedules, including extra sections (added nonscheduled flights) and related nonrevenue flights.

Secondary Functions. Those operations that are supplementary to a basic function. For example, for a five-cent lead pencil, the lead itself performs the basic function of writing and therefore is the basis of the original design, while the eraser and wood sheath perform secondary functions.

Supplemental Air Carrier. An air carrier certificated in accordance with FAR Part 121 and authorized to conduct nonscheduled or supplemental carriage of passengers or cargo or both in air transportation.

Supplemental (All Cargo). An air carrier certificated in accordance with FAR Part 121 and authorized to transport cargo by aircraft pursuant to approved and published schedules and routes.

Technical Standard Orders (TSO). FAA performance and quality control standards for certain materials, parts, and appliances such as avionics, seats, instruments, tires, etc.

Total Time on Airframe and Engines (TTAE). The total hours an aircraft and its engines have been in operation.

Turbofan Aircraft. Aircraft propelled by the turbojet engine whose thrust has been increased by the addition of a low-pressure compressor (fan). The turbofan engine can have an oversized low-pressure compressor at the front, with part of the flow bypassing the rest of the engine (front-fan or forward-fan), or it can have a separate fan driven by a turbine stage (aft-fan).

Turbojet Aircraft. Aircraft propelled by the jet engine incorporating a turbine-driven air compressor to take in and compress the air for the combustion of fuel, the gases of combustion (or the heated air) being used to rotate the turbine and create a thrust-producing jet.

Turboprop Aircraft. Aircraft in which the main propulsive force is supplied by the conventional propeller driven by a gas turbine. Additional propulsive force may be supplied from the discharge turbine exhaust gas.

Turboshaft Helicopter. A helicopter powered by one or more gas turbine engines.

Value. A measure of worth. Good value is obtained when the lowest possible price is paid to obtain reliable functions.

Value Analysis (VA). Same as Value Engineering.

Value Assurance. Value Engineering efforts conducted during the design and development stages of a product which assure good value in the product.

Value Engineering. A system of step-by-step procedures used to identify and eliminate elements of unnecessary cost which are not required to provide "customer required," essential characteristics of the product.

Value Engineering Change Proposal (VECP). A proposal which embodies a request to make a change to a product or service to reduce cost. In most cases, a VECP is a package of documentation which provides all data and information regarding a value engineering study with the recommended changes including estimates of the change impact on schedule, costs, performance, reliability, logistics, etc.

Value Improvement. Value Engineering actions conducted for released designs which improve the value characteristics of the product.

Wide-Body Aircraft. A generic and commonly used term applied to any and all of the newest generation of jet aircraft (turbofans) with a fuselage diameter exceeding 200 inches and whose per engine thrust is greater than 30,000 pounds (i.e., Boeing 747, McDonnell Douglas DC-10, Lockheed L-1011).

Notes

1. Introduction

1. *Air Transport, 1983: The Annual Report of the U.S. Scheduled Airline Industry* (Washington, D.C.: Air Transport Association of America, June 1983), p. 8.
2. *Air Transport, 1984: The Annual Report of the U.S. Scheduled Airline Industry* (Washington, D.C.: Air Transport Association of America, June 1984), p. 7.
3. *The General Aviation Story* (Washington, D.C.: General Aviation Manufacturers Association, 1979).
4. The Civil Aeronautics Board, in January 1981, reclassified airlines into four categories: majors (annual revenues of over $1 billion); nationals (annual revenues of $75 million to $1 billion); large regionals (annual revenues of $10 million to $75 million); and medium regionals (annual revenues of less than $10 million).
5. David A. NewMyer, "The Aviation Industry: An Employment Outlook," *Journal of Studies in Technical Careers*, vol. 3, no. 3 (Summer 1981).
6. "Safety-Rate of Carrier Accidents Reducted Sharply in 1980," *Aivation Week and Space Technology*, Jan. 1981. p. 80.
7. National Transportation Safety Board, *Annual Report to Congress, 1983* (Washington, D.C.: NTSB, June 1984), p. 9.
8. Most of this background section of the test is taken from an excellent article (published in pamphlet form) by Robert J. Serling, *Wrights to Wide-Bodies: The First Seventy-Five Years* (Washington, D.C.: Air Transport Association of America, 1978).
9. Ibid., p. 20.

10. C. N. VanDeventer, *An Introduction to General Aeronautics,* 3rd ed. (American Technical Society, 1974), p. 20.

2. THE FEDERAL AVIATION ADMINISTRATION

1. Source for description of the FAA: U.S. Department of Transportation, FAA, "The Federal Aviation Administration."
2. U.S. Department of Transportation, FAA, "The FAA Organization," Order 1100.5A, Change 65, Aug. 19, 1981, p. 202.
3. The Boeing Co., "Maintaining Continuous Aircraft Airworthiness" (Unpublished report, The Boeing Co., Seattle, Wash., Jan. 22, 1980).
4. Courtesy of Beech Aircraft Corporation, Wichita, Kans., 1983.
5. Source: U.S. Department of Transportation, FAA, "FAA Organization—Policies and Standards," Order 1100.1A, Change 1, Jan. 25, 1982, p. 2.
6. Ibid., p. 1.
7. U.S. Department of Transportation, FAA, "Air Carrier Airworthiness Inspector's Handbook," Order 8320.12, Change 2, June 28, 1978, p. 595.

3. REGULATORY REQUIREMENTS

1. U.S. Department of Transportation, FAA, FAA-APA-PG-5, "Guide to Federal Aviation Administration Publications," June 1982, p. 1.
2. U.S. Department of Transportation, FAA, Advisory Circular, AC-00-2VV, Apr. 15, 1982.
3. "Aircraft Maintenance Responsibilities" (Unpublished guidance material compiled by District Office Inspectors, 1982).
4. Source: Much of the information contained in this section is through the courtesy of The Boeing Company and is contained in its "Maintaining Continuous Aircraft Airworthiness" (Unpublished report, Seattle, Wash., Jan. 22, 1980).
5. *Air Transport, 1982: The Annual Report of the U.S. Scheduled Airline Industry* (Washington, D.C.: Air Transport Association of America, June 1982), p. 9.
6. The Boeing Co., "Maintaining Continuous Aircraft Airworthiness."
7. Ibid.
8. Methods of repair are FAA-approved.

4. ORGANIZATIONAL STRUCTURES

1. Some organizational charts omit the staff alignments.
2. Douglas Aircraft Co., "DC-9 Maintenance Facility and Equipment Planning Manual" (Report 761-145, Long Beach, Calif., June 1982), Table 3.1, p. 111-07.
3. Source of information about United Airlines: Courtesy of United Airlines, News Release, San Francisco, Calif., 1983.

5. Management Responsibilities

1. The functions of management are often broken down differently, depending on whose textbook is being used. Some texts omit the staff function; then there are other texts that go further by including such major areas as representing and innovating.

2. Raymond E. Glos et al., *Business: Its Nature and Environment*, 8th ed., (Cincinnati, Ohio: Southeastern Publishing Co., 1976), p. 124.

3. Ibid.

4. "An example of a strategic decision was Lockheed's decision in 1968 to compete against the McDonnell-Douglas DC-10 with its L-1011 'Airbus.' This required an enormous marshalling of persons, materials, machines, and money. Over $40 million had to be committed to new buildings alone, before a single L-1011 could be built" (Glos et al., *Business*, pp. 125-26).

5. See "Maintenance Facility Planning," by J. M. Leavens, Ground Support Equipment and Maintenance Facilities, McDonnell Douglas Corp., Long Beach, Calif., *Douglas Service*, Vol. 39, Mar./Apr. 1982.

6. John E. Heintzelman, *The Complete Handbook of Maintenance Management*, 5th ed. (Englewood Cliffs, N.J.: Prentice-Hall, 1980), p. 101.

7. Ibid., p. 103.

8. Ibid., pp. 106–12.

9. Ibid., pp. 162–65.

10. L. R. Bittel, *What Every Supervisor Should Know*, 4th ed. (New York: McGraw-Hill Book Co., 1980), pp. 170–71.

11. J. D. Richardson, *Essentials of Aviation Management* (Dubuque, Iowa: Kendall/Hunt Publishing Co., 1977). p. 22.

12. *Webster's New Collegiate Dictionary* (Springfield, Mass.: G. & C. Merriam Co., 1981), p. 648.

13. The seminar is just one step Bell has taken to assist its supervisors in developing a better understanding of worker needs.

14. Glos et al., *Business*, p. 135.

15 *Harvard Business Review* on management, 1st ed. (Cambridge: Harvard Business School, 1981), p. 22.

16. Usually developed through industrial engineering motion and time study methods.

17. Glos et al., *Business*, p. 124.

6. Aviation Maintenance Procedures

1. U.S. Department of Transportation, FAA, "Continuous Airworthiness Maintenance Programs," Advisory Circular 120-16C, Aug. 8, 1980.

2. These three parts pertain to maintenance and inspection programs of air carriers and commercial operators under FAR Parts 121 and 127 and operators of aircraft type certificated for 10 or more passenger seats under Part 135.

3. U.S. Department of Transportation, FAA, "General Aviation Airworthiness Inspector's Handbook," Order 8600.1, Oct. 16, 1978, p. 161.

4. The manual should accommodate work performed for the certificate holder by other persons. The policies and procedures segment of the manual should assign responsibilities and delineate procedures for the administrative aspect of contracted work.

5. Authority: The regulatory basis for maintenance facility inspections is contained in FAR Sections 121.81, 121.105, 121.123, 121.363, 121.365, 121.367, 121.369, 127.29, 127.132, 127.133, 127.134, and 127.135 (Order 8320.12 Chapter 3, Paragraph 480, Nov. 7, 1977).

6. Authority: The regulatory basis for specifications governing maintenance of aircraft (Operations & Specifications) is contained in Federal Aviation Regulations (FAR) 121.25, 121.45, 127.13, and 135.5.

7. Progressive Inspection: A progressive inspection is an inspection system designed to break a 100-hour or annual inspection down into small parts capable of completion on an hourly or calendar time basis. The purpose for using this type of inspection is to keep an aircraft from being out of commission for a long period of time while the complete 100-hour or annual inspection is being performed and keep it in airworthy condition at all times.

8. ATCO is an acronym for Air Taxi Commercial Operator.

9. U.S. Department of Transportation, FAA, "Air Carrier Airworthiness Inspector's Handbook," Order 8320.12, Change 6, Section 17, pp. 709–12.

10. Ibid., pp. 1301, 1303, 1333.

7. APPLICATIONS OF AVIATION MAINTENANCE CONCEPTS

1. A repair station located outside of the United States. For the reader's interest: an accident occurred in 1982 that involved a foreign firm holding an approved United States repair certificate. Spanish government officials investigated the crash of a Spantax McDonnell Douglas DC-10 that occurred at Malaga airport on September 13, 1982. Focus of the DC-10 inquiry was on the failure of a retreaded nose gear tire. "All we know of the Spanish investigation is that the retread was done by a Swiss firm holding an approved U.S. repair certificate," an FAA official said. (Source: Article in *Aviation Week & Space Technology*, Oct. 4, 1982, p. 33.)

2. The repairman, so certified, must be at or above the level of shop foreman or department head (Paragraph 145.41).

3. J. D. Richardson, *Essentials of Aviation Management*, (Dubuque, Iowa: Kendall/Hunt Publishing Co., 1977), p. 311.

4. Ibid.

5. Source: Part 91, "General Operating and Flight Rules," Subpart C—Maintenance, Preventive Maintenance, and Alterations.

6. Source: Order 8600.1, "General Aviation Airworthiness Inspector's Handbook," Section 15, Chapter 2, Oct. 16, 1978, p. 171.

7. If the aircraft is used for hire, it must have an inspection every 100 hours. The only difference between the 100-hour inspection and the annual inspection is that the annual has to be signed by the holder of an inspection authorization whereas the 100-hour inspection has to be signed by a certified airframe and powerplant (A&P)

mechanic. The interval for the 100-hour inspection may be exceeded by 10 hours but those 10 hours must be deducted from the next 100-hour inspection. The annual inspection can count for a 100-hour inspection but the 100-hour inspection cannot count as an annual inspection.

8. In the *Air Transport, 1982,* put out by the Air Transport Association of America, Frontier Airlines ranked thirteenth out of 30 United States air carriers in passenger loadings (6,286,000). Although in the mid-bracket passengerwise for the last several years, Frontier has shown a strong financial return (net profit).

9. It takes approximately eight years of flying time to reach 22,500 flight hours.

8. BUDGETING, COST CONTROLS, AND COST REDUCTION

1. E. T. Newbrough, *Effective Maintenance Management* (New York: McGraw-Hill Book Co, 1967), p. 77.

2. John E. Heintzelman, *The Complete Handbook of Maintenance Management,* 5th ed. (Englewood Cliffs, N.J.: Prentice-Hall, 1980), p. 43.

3. L. D. Miles, *Techniques of Value Analysis and Engineering* (New York: McGraw-Hill Book Co., 1961), p. 1.

4. This section of the text based, by permission of the author, on C. W. (Smokey) Doyle's unpublished paper, "The Value Engineering Techniques." Mr. Doyle is the former Director of Cost Reduction and Value Control for General Dynamics, Ft. Worth, Texas.

5. Newbrough, *Effective Maintenance Management,* p. 327.

9. TRAINING AND PROFESSIONAL DEVELOPMENT IN AVIATION MAINTENANCE

1. The National Microfilm Library, *The College Catalog Collection,* Card OK 16, p. 6.

2. Ibid., p. 7.

3. U. S. Department of Transportation, FAA, "Aviation Mechanic General Written Test Guide," EA-AC 65-20A, 1981, Forword.

4. Ibid., p. 1.

5. U.S. Department of Transportation, FAA, Office of Airworthiness, "Mechanics Certification Guide," AC 65-2D, revised 1983, pp. 10, 11.

6. Ibid.

7. U.S. Department of Transportation, FAA, "Aviation Mechanic General Written Test Guide," EA-AC 65-20A, 1984, p. 14.

8. U.S. Department of Transportation, FAA Office of Airworthiness, "Mechanics Certification Guide," AC 65-2D, revised 1983, p. 11.

9. U. S. Department of Transportation, FAA, "Written Test Answers and Explanations, Powerplant Mechanics" EA-AC 65-224, 1979 (Basin, Wyo.: Aviation Maintenance Publishers, 1979).

10. U.S. Department of Transportation, FAA, Office of Airworthiness, 'Mechanics Certification Guide," AC 65-2D, revised 1983, p. 10.

11. The National Microfilm Library, *The College Catalog Collection,* Card OK 16, p. 6.

12. *Federal Aviation Regulations* (U.S. Government Printing Office, 1982), Part 65.87, p. 9.

13. *Federal Aviation Regulations,* Part 65.85, p. 9.

14. Material in this section of the text courtesy of Colorado Aero Tech, Inc., Broomfield, Colorado, 1983.

10. SAFETY AND MAINTENANCE

1. Most of the first part of this chapter is taken from an excellent article by Robert J. Serling, *Wrights to Wide-Bodies: The First Seventy-Five Years* (Washington, D.C.: Air Transport Association of America, 1978).

2. "Safety-Rate of Carrier Accidents Reduced Sharply in 1980," *Aviation Week and Space Technology,* Jan. 1981, p. 80.

3. Two recent special studies developed by NTSB were (1) special investigation report on "Air Traffic Control System," Dec. 8, 1981, and (2) special study on "Cabin Safety in Large Transport Aircraft," Sept. 9, 1981.

4. Newspaper account taken from the *Times-Union and Journal,* Jacksonville, Fla., Jan. 24, 1982.

5. F. H. King, "Impact of Unsafe Aviation Ground Operations: United States Scheduled Airlines and Selected Foreign Carriers" (Ph.D. diss., Southern Illinois Univ., 1981), p. 1.

6. "Maintenance Man Ingested by Jet," *Shreveport Times,* Shreveport, La., Mar. 3, 1981, p. 27.

7. "AA Accident Prevention Checklist," Safety Division, American Airlines, Inc., Jan. 1980, pp. 1-2.

8. AA Safety Division, "AA Accident/Incident Report," 1978.

9. A questionnaire was developed by the author on aircraft accidents and costs. In the responses received regarding accident costs and workmen's compensation, six airlines indicated a total accident cost of $7,227,000 and workmen's compensation costs of $24,822,000 in 1981.

10. Except for the numbered footnote references, all material relating to safety in the maintenance hangar and on the ramp was taken, in abbreviated form, from Air Force Regulation 127-101, Sept. 4, 1974, and FAA AC 150/52-30-4, Aug. 27, 1982.

11. National Safety Council, *Aviation Ground Operation—Safety Handbook,* 3rd ed. (NSC, Chicago, 1977), p. 51.

12. Douglas Aircraft Co., "DC-9 Maintenance Facility and Equipment Planning Manual" (Report 761-145, Long Beach, Calif., June 1982), p. 71-15.

13. FAA Advisory Circular, AC 150/5230-4, "Aircraft Fuel Storage, Handling and Dispensing on Airports," U.S. Department of Transportation, Aug. 27, 1982.

14. Douglas Aircraft Co., "DC-9 Maintenance Facility and Equipment Planning Manual," p. 12-03.

15. National Safety Council, *Aviation Ground Operation—Safety Handbook*, p. 62.
16. Ibid, p. 65.

11. Electronic Data Processing

1. CAMP/DISPATCH material used courtesy of Aviation Information Services, Inc.

12. Aviation Maintenance Management Problem Areas

1. "Think Parts, Think Sales, Think Profits," *Airport Services Management* (Lockwood Publications), Apr. 1982, p. 48.
2. Ibid.
3. Ibid., p. 50.
4. Ibid.
5. Material on the FAA maintenance inspector taken from: "The FAA Maintenance Inspector," *Airport Services Management* (Lockwood Publications), Dec. 1981.
6. "10 Ways to Ruin Your Business," *Airport Services Management* (Lockwood Publications), Apr. 1982, p. 50.

13. Forecast and Summary

1. Data source for this chapter is the U.S. Department of Transportation, FAA, *FAA Aviation Forecasts: Fiscal Years 1980–1991*, Sept. 1979.
2. Ibid., p. 32.
3. Department of Transportation, FAA, Air Transportation Regulation, *Federal Register*, Vol. 47, No. 182, Mon., Sept. 20, 1982, Proposed Rules, p. 41487.
4. Source: *Air Transport, 1982: The Annual Report of The U.S. Scheduled Airline Industry* (Washington, D.C.: Air Transport Association of America, June 1982), p. 1.

Index

A & P certificate, 4, 5, 115–17, 120
Accidents, at repair stations: at battery shop, 143–44; electrical accidents, 140–41, 143; at paint shop, 143; preventive maintenance, 144; radiation accidents, 141–43
Accidents and fatalities, on aircraft: on air taxis, 6–7; avoidance of, 20; causes of, 17, 20, 124; on commuter air carriers, 6; compared to auto fatalities, 123; by crash landing, 127; examples of, 6, 123–24, 127; on general aviation airlines, 5–6, 7, 126; investigation of, 124–25; rates of, 6–7; on scheduled air carriers, 5–6. *See also* Safety
Administrative paperwork problems, 158
Advisory Circulars (ACs), 18, 33–34, 94; partial listing of, 167–75
Aeromedical research, 20, 21
Aeronautics Branch, Department of Commerce, 12
Air cargo, 2, 8, 191; on United Airlines, 53, 55
Air Commerce Act of 1926, 12
Aircraft: FAA certification of, 14–17, 32, 37–42, 43–44; inspection and overhauls of, 17, 24–27, 37–41, 79–81, 86–87, 123; maintenance programs for new models, 44–45, 46; manufacturers of, 7, 17, 42, 43–44, 46, 48; numbers of, 2, 10, 11; publications on maintenance of, 47, 81–82; reliability programs for, 25, 28; small, 87, 88–89; surveillance of, 28
Aircraft Utilization and Propulsion Reliability Report, 33
Air flight, history of, 8–10, 12
Airframe technicians, education of, 118
Airline Deregulation Act of 1978, 161
Airlines. *See* General aviation airlines; Scheduled airlines; *names of individual airlines*
Airport and Airway Development Act of 1970, 17, 18
Airport Development Aid Program (ADAP), 17
Airport Improvement Program, 17–18
Airports: FAA operation of, 21; improvements to, 17–18; number of, 2
Air taxis: accidents on, 6–7; definition of, 189; FAA certification of, 33; maintenance on, 42, 99; repair station inspections, 84–85; and safety, 124
Air traffic control, by FAA, 13, 191; automation of, 14, 20
Air traffic controllers, 17
Air travel clubs, 99, 191

Air turbulence, danger of, 20
Airworthiness: Boeing approach toward, 42, 45; definition of, 79; FAA certification of, 15; FARs for, 35, 78–79, 99; listing of requirements and forms for, 177–89; and quality assurance, 156–57. *See also* Airworthiness Directives; General Aviation Airworthiness Alerts
Airworthiness Directives (ADs), 32, 93, 125
American Airlines, safety program at, 127–29
Anti-hijacking Act of 1974, 19
Aviation gasoline (AVGAS), 138–40, 192. *See also* Fuel
Aviation industry, types of, 2
Aviation maintenance, importance of, 1
Aviation maintenance technicians, 117
Avionics technicians, 118

Beech Aircraft Corporation, 16–17
Bird hazards, 18
Boeing Air Transport, 54
Boeing Company: aircraft, 42, 43–44; maintenance development by, 47; organization of, 52
Budgets: development of, 106–7; elements of, 108; problems in, 108; purpose of, 106, 107, 113
Bureau of Air Commerce, 12

Capital Airlines, 54
Certification, by FAA: of air carriers, 192; of aircraft, 14–17, 32, 37–42, 43–44, 87–88, 94; of air taxis, 33; authority granted for, 12, 13; criteria for, 5, 36; of personnel, 14, 17, 33, 36, 93, 96–97, 114; of products and parts, 93; of repair stations, 17, 36–37, 94–96, 157; and safety, 17, 124; of technical schools, 4, 42
Charter air carriers, 192
Civil Aeronautics Authority, 12
Civil Aeronautics Board (CAB), 12
Civil Aviation Assistance Groups, 21
Civil Aviation Security Program, 13, 18
Clean Air Act of 1970, 19
Code of Federal Regulations (CFRs), 32
Collision avoidance systems, 20
Colorado Aero Tech, 119–20
Communications: between personnel, 73–74; plane-to-ground, 53
Commuter airlines: accidents on, 6; definition of, 192; expansion of, 3, 162; maintenance programs of, 99; organization of, 64–65; and safety, 124
Computers, 148, 149. *See also* Electronic data processing
"Condition-monitoring" maintenance, 27, 28

Cost controls, 108, 153–56
Cost reduction: definition of, 192; types of, 108–9; value analysis (VA) for, 108–12
Crash landings. *See* Accidents and fatalities, on aircraft

Damage tolerance requirements, 25, 27
"Decision tree analysis" maintenance, 25–26
Defueling. *See* Fueling procedures
Direct maintenance expenses, 192–93
Discrete Address Beacon System (DABS), 20
Dispatchers, 17
Douglas Aircraft Company, 54
Dulles International Airport, 21

Education and training schools: admissions to, 114–15; examples of, 119–21; FAA certification of, 17, 114, 115–17; FAA regulation of, 4–5, 34, 42, 93; for managers, 4–5, 96; for pilots, 121; programs in, 117, 118; publications on, 34
Electronic data processing: advantages of, 143; specialty systems for, 145–46, 149–50; types of data stored, 146, 150–51
Engineering and Development (E & D) projects, 19–20
Engines: inspection of, 58, 62, 81; manufacturers of, 54; safe handling of, 131; test cells for, 57, 59
Environmental protection, by FAA, 13, 18–19

"FAA Organization" (manual), 22
FAA regulations (FARs), 24, 31; on aircraft inspection, 37–41; on air taxis, 42; airworthiness responsibilities defined in, 35; contents of, 34–35; definition of, 193; development of, 163–64; on maintenance records required, 39–41; maintenance responsibilities in, 35, 36–37, 78, 82, 83, 84, 89, 90, 93–94, 100; on maintenance technical schools, 4, 42; on personnel certification, 36; on repair stations, 36–37, 93
Fares, 3
Fatalities. *See* Accidents and fatalities, on aircraft
Federal Airport Act of 1946, 17
Federal Aviation Act of 1958, 12
Federal Aviation Administration (FAA): airports operated by, 21; approval by, 34; creation of, 12–13; engineering and development projects of, 19–20; environmental protection by, 13, 18–19; nonregulatory and support publications of, 30–31, 33–34; organizational manual of, 22; regional boundaries of, 22–23; regulatory and technical publications of, 30, 31–33; responsibil-

Index

ities of, 13, 30; weather information from, 19. *See also* Certification, by FAA
Federal Aviation Agency, 12–13
Finances. *See* Budgets; Cost controls; Value analysis
Fire-fighting equipment, 18
Fixed base operators (FBOs): definition of, 193; maintenance by, 100–105; organization of, 65, 66
Flight engineers, 17
Flight standards forms, 32–33
Freight. *See* Air cargo
Frontier Airlines: maintenance by, 100, 101; organization of, 51; repair stations of, 90, 91–92. *See also* Colorado Aero Tech
Fuel: cost of, 3, 25; efficiency of, 162; explosions of during crash landings, 126–27; fire prevention near, 130, 131–32; toxicity of, 131. *See also* Aviation gasoline; Jet fuel
Fueling procedures, 134–36; defueling, 140; overwing servicing, 137; safety during, 134, 138; underwing servicing, 136–37
Fueling systems, 138
Fueling vehicles, 138
Functional organization structures, 50, 51

General aviation airlines: accidents on, 5–6, 7, 126; definition of, 193; growth of, 162; and safety, 124; size of, 2, 11
General Aviation Airworthiness Alerts, 15–16, 94
General Aviation Manufacturers Association (GAMA), 7
Governmental regulation, reduction of, 3. *See also* FAA regulations

"Hard-time" maintenance, 24–25, 27–28
Hardware failure, critical tests for, 26
Hazardous materials: handling of, 130–33; inspections for, 85
Helicopters, 193
Hijackings, 19
History, of air flight, 8–10, 12

Independent Safety Board Act, 125
Indirect maintenance expenses, 193
Inert gas welding, 59–60, 63
Inertial navigation equipment, 57, 58
Inspection: of aircraft, 17, 24–27, 37–41, 79–81, 123; of repair stations, 83–85, 86–87, 157–58
Inspectors, 97
Instrument flight rules (IFRs), 13
International Civil Aviation Organization (ICAO), 21

Jacking, safety during, 131, 133–34
Jet fuel, 138, 139, 140, 194. *See also* Fuel

Lindbergh, Charles, 12
Line and staff organization structure, 50
Line organization structure, 50, 51
Load factor, 194
Logbooks, inspection of, 86

Maintenance bulletins, 97–98
Maintenance manuals: contents of, 81–82; inspections of, 86; organization outline of, 89–90; types of, 16–17
Maintenance Operations Center (MOC), United Airlines: description of, 55, 58; facilities of, 56–60; organization of, 55, 56, 61–64; statistics on, 60
Maintenance processes, primary, 27–29; responsibilities specified in FARs, 35, 36–37, 78, 82, 83, 84, 89, 90, 93–94, 100; on small aircraft, 88–89
Maintenance records required by FAA, 39–41, 99
Maintenance Review Board (MRB), 45, 46
Maintenance Steering Group Revision No. 3 (MSG-3), 25–27
Maintenance Steering Groups (MSGs), 45
Maintenance technician schools. *See* Education and training schools
Malfunction or Defect Reports (FAA Form 8010-4), 16
Managers: characteristics of, 4; controlling by, 74–77; directing by, 67, 72–74; education and certification for, 4–5, 96; major failings of, 159; organizing by, 67, 69–70; planning by, 67, 69–70; problems confronted by, 152; role of, 1, 4, 67; staffing by, 67, 70–72, 96
Manpower. *See* Personnel
Manuals. *See* Maintenance manuals
Mechanics: FAA certification of, 17, 114; shortages of, 7, 162
Microwave Landing System (MLS), 20–21
Monroney Aeronautical Center, 21

National Air Transport, 54
National air transportation system: growth of, 160–61, 164; impact on industry of, 8–9; role of, 3, 8
National Transportation Safety Board (NTSB), 21, 125
Navigators, 17
Noise, reduction of, 17–18
Notices of Proposed Rulemaking (NPRMs), 31

"On-condition" maintenance, 24–25, 28
Organizational structures, types of, 50–52

Pacific Air Transport, 54
Parachute riggers, 17, 34
Parts, FAA certification of, 93
Parts department, 153
Passengers: early service for, 53; on scheduled air carriers, 4, 8, 9; screening of, 19
Performance criteria tests, 5
Personnel: in air traffic control, 13, 14; FAA certification of, 14, 17, 33, 36, 93, 96–97, 114; incentives for, 75–77; injury to, 130–31; motivation for, 72–73; numbers of, 2; shortages of, 7, 160, 162; staffing by managers, 67, 70–72; 96; at United Airlines, 55
Pilots, 17, 123
Planning Grant Program (PGP), 17, 18
Powerplant technicians, 117
Pratt and Whitney (engine manufacturers), 54
Production certificate, 43

Ramp inspections, 85, 86–87
Regional airlines, 3
Regional airports, 18
Regulation by Objective (RBO), 162
Regulatory Flexibility Act of 1980, 162
Reliability programs, for aircraft, 25, 28; reports on, 33
Repair stations: accidents at, 141–44; FAA approval of, 34; FAA certification of, 17, 36–37, 94–96, 157; FARs on, 36–37, 93; inspections of, 83–85, 157–58; personnel of, 96
Rescue equipment, 18
Rulemaking materials, 31
Runways, marking of, 18

Safety: at American Airlines, 127–29; and crime prevention, 19; during fueling, 134, 138; experiments for, 126–27; FAA standards for, 18, 21; and fire prevention, 18, 130, 131–32; as goal of FAA certification, 17, 124; Independent Safety Board Act, 125; in the maintenance hangar, 129–34; National Transportation Safety Board (NTSB), 21, 125; for personnel, 130; priority for, 122–23; in terminals, 19
Scheduled airlines: accidents on, 5–6; passengers on, 4, 8, 9, 19, 53; reduction of, 3; size of, 2, 11. See also Commuter airlines; Regional airlines
Service difficulty reports, 33
Service support, by aircraft manufacturers, 17
Sonic booms, 18
Southern Illinois University, aviation curriculum at, 120–21
Spot inspections, 85–86
Supplemental type certificates (STCs), 94

Taxiways, 18
Technical schools. See Education and training schools
Technical Standard Orders (TSO), 33, 94, 195
Training. See Education and training schools
Trunk carriers, 193. See also Scheduled airlines
Turbofan aircraft, 194
Turbojet aircraft, 195
Turboprop aircraft, 162
Type certifications: for aircraft, 32, 43–44, 94; for small aircraft, 87–88

United Airlines: aircraft fleet of, 53; employees of, 55; facilities of, 55; origins of, 54; route system of, 53, 55. See also Maintenance Operations Center (MOC), United Airlines

Value analysis (VA), 108–12, 196
Varney Air Lines, 54

Washington National Airport, 21
Weather information from FAA, 19
Wind shear, danger of, 20
Wright Brothers, 8

FRANK H. KING, Ph.D, P.E., holds a B.S.M.E. degree from Virginia Polytechnic Institute and State University, an M.B.A. degree from International University, an M.S.I.E. degree from Oklahoma State University, and a Ph.D. degree in education from Southern Illinois University. Dr. King is retired from the United States Air Force and Civil Service, having held positions as project engineer, project manager, and lecturer on ballistic missiles and space programs. He has been selected several times for inclusion in *Who's Who in America*. He has been an industrial engineering and aviation consultant and has conducted numerous lectures throughout the United States in both fields. Dr. King is currently teaching, on an adjunct basis, an aviation course for Southern Illinois University.